Collins

Cambridge International
AS & A Level Mathematics

Probability & Statistics 1

STUDENT'S BOOK

Louise Ackroyd, Jonny Griffiths, Yimeng Gu
Series Editor: Dr Adam Boddison

William Collins' dream of knowledge for all began with the publication of his first book in 1819.

A self-educated mill worker, he not only enriched millions of lives, but also founded a flourishing publishing house. Today, staying true to this spirit, Collins books are packed with inspiration, innovation and practical expertise. They place you at the centre of a world of possibility and give you exactly what you need to explore it.

Collins. Freedom to teach.

Published by Collins
An imprint of HarperCollins*Publishers*
The News Building
1 London Bridge Street
London
SE1 9GF

HarperCollins*Publishers* Macken House
39/40 Mayor Street Upper
Dublin 1
DO1 C9W8
Ireland

Browse the complete Collins catalogue at
www.collins.co.uk

© HarperCollins*Publishers* Limited 2018

10 9 8 7 6

978-0-00-825776-7

All rights reserved. No part of this publication may be reproduced, stored in a retrieval system, or transmitted in any form by any means, electronic, mechanical, photocopying, recording or otherwise, without the prior written permission of the Publisher or a licence permitting restricted copying in the United Kingdom issued by the Copyright Licensing Agency Ltd., Barnard's Inn, 86 Fetter Lane, London, EC4A 1EN.

British Library Cataloguing in Publication Data

A catalogue record for this publication is available from the British Library.

Commissioning editor: Jennifer Hall
In-house editor: Lara McMurray
Authors: Louise Ackroyd/Jonny Griffiths/Yimeng Gu
Series editor: Dr Adam Boddison
Development editor: Tim Major
Project manager: Emily Hooton
Copyeditor: Dr Jan Schubert
Proofreaders: Gwynneth Drabble/Gudrun Kaiser
Reviewer: Adele Searle
Answer checkers: David Hemsley/Marie Taylor
Cover designer: Gordon MacGilp
Cover illustrator: Maria Herbert-Liew
Typesetter: Jouve India Private Ltd
Illustrators: Ken Vail Graphic Design Ltd/Jouve India Private Ltd
Production controller: Sarah Burke
Printed and bound by Ashford Colour Press Ltd
®IGCSE is a registered trademark

This book contains FSC™ certified paper and other controlled sources to ensure responsible forest management.

For more information visit: www.harpercollins.co.uk/green

Acknowledgements

The publishers wish to thank Cambridge Assessment International Education for permission to reproduce questions from past IGCSE® Mathematics and AS & A Level Mathematics papers. Cambridge Assessment International Education bears no responsibility for the example answers to questions taken from its past papers. These have been written by the authors. Exam-style questions and sample answers have been written by the authors. The publishers wish to thank the following for permission to reproduce photographs. Every effort has been made to trace copyright holders and to obtain their permission for the use of copyright material. The publishers will gladly receive any information enabling them to rectify any error or omission at the first opportunity.

pvi LatinStock Collection/Alamy Stock Photo, p1 LatinStock Collection/Alamy Stock Photo, p44 Luis Alvarez/DigitalVision/Getty images, p83 Prasit Rodphan/Alamy Stock Photo, p117 Pavel L Photo and Video/Shutterstock.

Full worked solutions for all exercises, exam-style questions and past paper questions in this book available to teachers by emailing international.schools@harpercollins.co.uk and stating the book title.

CONTENTS

Full worked solutions for all exercises, exam-style questions and past paper questions in this book available to teachers by emailing international.schools@harpercollins.co.uk and stating the book title.

INTRODUCTION

This book is part of a series of nine books designed to cover the content of the Cambridge International AS & A Level Mathematics and Further Mathematics. The chapters within each book have been written to mirror the syllabus, with a focus on exploring how the mathematics is relevant to a range of different careers or to further study. This theme of *Mathematics in life and work* runs throughout the series with regular opportunities to deepen your knowledge through group discussion and exploring real-world contexts.

Within each chapter, examples are used to introduce important concepts and practice questions are provided to help you to achieve mastery. Developing skills in modelling, problem solving and mathematical communication can significantly strengthen overall mathematical ability. The practice questions in every chapter have been written with this in mind and selected questions include symbols to indicate which of these underlying skills are being developed. Exam-style questions are included at the end of each chapter and a bank of practice questions including real Cambridge past exam questions are included at the end of the book.

A range of other features throughout the series will help to optimise your learning. These include:

> **key information boxes** – highlighting important learning points or key formulae

> **commentary boxes** – tackling potential misconceptions and strengthening understanding through probing questions

> **stop and think** – encouraging independent thinking and developing reflective practice.

Key mathematical terminology is listed at the beginning of each chapter and a glossary is provided at the end of each book. Similarly, a summary of key points and key formulae is provided at the end of each chapter. Where appropriate, alternative solutions are included within the worked solutions to encourage you to consider different approaches to solving problems.

Probability & Statistics 1 will introduce the concepts required to make informed predictions about the outcomes of real-life events. You will learn how to analyse data sets by comparing measures of central tendency alongside the spread of the data and you will understand more about the validity and appropriateness of different statistical representations. You will see how aspects of pure mathematics, such as the binomial theorem, can be applied in statistical contexts.

This book will demonstrate the value of statistics in a variety of different careers, ranging from quality control officers working in a factory to actuaries making predictions about the future. One of the most well-known statistical distributions, the normal distribution, will be used to model a range of real-life scenarios, including how doctors identify babies who are significantly underweight and claims of the reliability of success made by manufacturers of new products.

FEATURES TO HELP YOU LEARN

Mathematics in life and work

Each chapter starts with real-life applications of the mathematics you are learning in the chapter to a range of careers. This theme is picked up in group discussion activities throughout the chapter.

Learning objectives

A summary of the concepts, ideas and techniques that you will meet in the chapter.

Language of mathematics

Discover the key mathematical terminology you will meet in this chapter. Throughout the chapter, key words are written in bold. These words are defined in the glossary at the back of the book.

Prerequisite knowledge

See what mathematics you should know before you start the chapter, with some practice questions to check your understanding.

Explanations and examples

Each section begins with an explanation and one or more worked examples, with commentary where appropriate to help you follow. Some show alternative solutions in the example or accompanying commentary to get you thinking about different approaches to a problem.

---

1 REPRESENTATION OF DATA

Mathematics in life and work

You are surrounded by data. Almost every news bulletin contains numerical data presented in a variety of ways. More and more people collect data from you, often without you realising it. Computers are able to process data about every aspect of your life more and more quickly, to be used by people you will never meet. Even online advertisements are selected based on data that has been gathered about your purchasing habits. Data and how it is collected have never been so heavily scrutinised.

Data representation can be involved in a variety of careers. For example:

> If you were a journalist trying to hold the government to account, you might need to determine the truth about how much money is being spent in a particular department, and how.

> If you were an actuary working with data on life expectancy, you might need to isolate key parts of the data about a person or a population.

> If you were a doctor researching the spread of a disease, you might need to decide on the most truthful way of conveying your data. Your choice might affect millions of people: how can you show the danger most clearly?

This chapter includes some of the problems you might encounter if you were a doctor.

LEARNING OBJECTIVES

You will learn how to:

> choose suitable ways of presenting qualitative and quantitative raw data, discussing the advantages and disadvantages of your choice

> use discrete, continuous, grouped and ungrouped data

> interpret, draw and use stem-and-leaf diagrams, histograms, box-and-whisker plots (including outliers) and cumulative frequency diagrams

> calculate and use measures of central tendency: mean, median and mode

> calculate and use measures of variation: range, interquartile range and standard deviation

> work with grouped and ungrouped data when calculating the mean and standard deviation.

LANGUAGE OF MATHEMATICS

Key words and phrases you will meet in this chapter:

> bimodal, box-and-whisker plot, categorical data coding, continuous data, cumulative frequency, discrete data, histogram, interquartile range, mean, median, mode, numerical data, outlier, percentile, qualitative data, quantitative data, quartile, range, standard deviation, stem-and-leaf diagram, variance

---

Example 3

Find the standard deviation for the data set {1, 4, 5, 6, 14} using both formulae. Comment on the relative ease of use of each formula.

Solution

Using the first formula:

$$\sum (x - \bar{x})^2 = (1 - 6)^2 + (4 - 6)^2 + (5 - 6)^2 + (6 - 6)^2 + (14 - 6)^2$$
$$= (-5)^2 + (-2)^2 + (-1)^2 + 0^2 + 8^2 = 94$$

and so s.d.$(x) = \sqrt{\dfrac{94}{5}} = 4.34 \,(3\,\text{s.f.})$

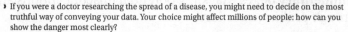 Most scientific calculators will allow you to input the data item by item and then calculate whatever statistics you require.

Using the second formula:

$$\sum x^2 = 1 + 16 + 25 + 36 + 196 = 274$$

vi

Colour-coded questions

Questions are colour-coded (green, blue and red) to show you how difficult they are. Exercises start with more accessible (green) questions and then progress through intermediate (blue) questions to more challenging (red) questions.

17 Evan has eggs every day, either scrambled (35% of the time) or poached (40% of the time) or fried (25% of the time). If his eggs are scrambled, he has them on toast 70% of the time, while poached eggs are on toast 75% of the time and fried eggs are on toast 25% of the time.

a Draw a tree diagram to show the possibilities.

b What is the probability that he has his eggs on toast?

c What is the probability that he chose fried eggs given that they were on toast?

18 A red die is rolled with result R, and a green die is rolled with result G. The results are multiplied together to obtain RG.

Question-type indicators

The key concepts of problem solving, communication and mathematical modelling underpin your A level Mathematics course. You will meet them in your learning throughout this book and they underpin the exercises and exam-style questions. All Mathematics questions will include one or more of the key concepts in different combinations. We have labelled selected questions that are especially suited to developing one or more of these key skills with these icons:

Problem solving – mathematics is fundamentally about problem solving and representing systems and models in different ways. These include: algebra, geometrical techniques, calculus, mechanical models and statistical methods. This icon indicates questions designed to develop your problem-solving skills. You will need to think carefully about what knowledge, skills and techniques you need to apply to the problem to solve it efficiently.

These questions may require you to:

> use a multi-step strategy

> choose the most efficient method, or bring in mathematics from elsewhere in the curriculum

> look for anomalies in solutions

> generalise solutions to problems.

Communication – communication of steps in mathematical proof and problem solving needs to be clear and structured and use algebra and mathematical notation so that others can follow your line of reasoning. This icon indicates questions designed to develop your mathematical communication skills. You will need to structure your solution clearly, to show your reasoning and you may be asked to justify your conclusions.

These questions may require you to:

> use mathematics to demonstrate a line of argument

> use mathematical notation in your solution

> follow mathematical conventions to present your solution clearly

> justify why you have reached a conclusion.

Mathematical modelling – a variety of mathematical content areas and techniques may be needed to turn a real-world situation into something that can be interpreted through mathematics. This icon indicates questions designed to develop your mathematical modelling skills. You will need to think carefully about what assumptions you need to make to model the problem and how you can interpret the results to give predictions and information about the real world.

These questions may require you to:

> construct a mathematical model of a real-life situation, using a variety of techniques and mathematical concepts

> ❯ use your model to make predictions about the behaviour of mathematical systems

> ❯ make assumptions to simplify and solve a complex problem.

Key information

These boxes highlight information that you need to pay attention to and learn, such as key formulae and learning points

KEY INFORMATION

Range = $Q_4 - Q_0$

Interquartile range = $Q_3 - Q_1$

Stop and think — You can compare two data sets using a back-to-back stem-and-leaf diagram or two box-and-whisker plots side by side. Which aspects of the data are shown best by each diagram?

Stop and think

These boxes present you with probing questions and problems to help you to reflect on what you have been learning. They challenge you to think more widely and deeply about the mathematical concepts, to tackle misconceptions and, in some cases, to generalise beyond the syllabus. They can be a starting point for class discussions or independent research. You will need to think carefully about the question and come up with your own solution.

Mathematics in life and work: Group discussions give you the chance to apply the skills you have learned to a model of a real-life mathematical problem involving a career that uses mathematics. Your focus is on applying and practising the concepts, and coming up with your own solutions, as you would in the workplace. These tasks can be used for class discussions, group work or as an independent challenge.

Summary of key points

At the end of each chapter, there is a summary of key formulae and learning points.

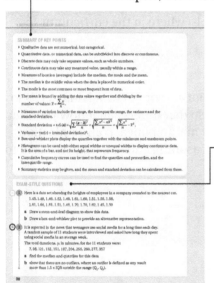

Exam-style questions

Practise what you have learned throughout the chapter with questions, written in examination style by our authors, that progress in order of difficulty.

The last **Mathematics in life and work** question draws together the skills that you have gained in this chapter and applies them to a simplified real-life scenario.

At the end of the book, test your mastery of what you have learned in the **Summary Review** section. Practise the basic skills and then go on some to try some carefully selected questions from Cambridge International A Level Mathematics questions and Further Mathematics past exam papers and exam-style questions. Extension questions, written by our authors, give you the opportunity to challenge yourself and prepare you for more advanced study.

1 REPRESENTATION OF DATA

Mathematics in life and work

You are surrounded by data. Almost every news bulletin contains numerical data presented in a variety of ways. More and more people collect data from you, often without you realising it. Computers are able to process data about every aspect of your life more and more quickly, to be used by people you will never meet. Even online advertisements are selected based on data that has been gathered about your purchasing habits. Data and how it is collected have never been so heavily scrutinised.

Data representation can be involved in a variety of careers. For example:

> If you were a journalist trying to hold the government to account, you might need to determine the truth about how much money is being spent in a particular department, and how.

> If you were an actuary working with data on life expectancy, you might need to isolate key parts of the data about a person or a population.

> If you were a doctor researching the spread of a disease, you might need to decide on the most truthful way of conveying your data. Your choice might affect millions of people: how can you show the danger most clearly?

This chapter includes some of the problems you might encounter if you were a doctor.

LEARNING OBJECTIVES

You will learn how to:

> choose suitable ways of presenting qualitative and quantitative raw data, discussing the advantages and disadvantages of your choice

> use discrete, continuous, grouped and ungrouped data

> interpret, draw and use stem-and-leaf diagrams, histograms, box-and-whisker plots (including outliers) and cumulative frequency diagrams

> calculate and use measures of central tendency: mean, median and mode

> calculate and use measures of variation: range, interquartile range and standard deviation

> work with grouped and ungrouped data when calculating the mean and standard deviation.

LANGUAGE OF MATHEMATICS

Key words and phrases you will meet in this chapter:

> bimodal, box-and-whisker plot, categorical data coding, continuous data, cumulative frequency, discrete data, histogram, interquartile range, mean, median, mode, numerical data, outlier, percentile, qualitative data, quantitative data, quartile, range, standard deviation, stem-and-leaf diagram, variance

PREREQUISITE KNOWLEDGE

You should already know how to:

 » use appropriate graphical representation for discrete, continuous and grouped data

 » interpret and construct tables, charts and diagrams, including frequency tables, bar charts, pie charts and pictograms for qualitative data, and vertical line charts for ungrouped discrete numerical data

 » construct and interpret histograms with equal class intervals and cumulative frequency graphs

 » choose an appropriate table, chart or diagram for a given situation

 » describe data as qualitative, quantitative, discrete or continuous as appropriate

 » use appropriate measures of averages and variation.

You may also know how to:

 » construct stem-and-leaf diagrams and box-and-whisker plots.

You should be able to complete the following questions correctly:

1   A test was given to 50 students and the following marks were awarded:

13	22	41	36	32	26	31	41	31	41
41	14	26	41	41	26	39	39	45	45
34	23	36	23	47	23	47	40	41	29
15	39	36	41	27	46	16	40	19	31
12	27	39	27	28	41	47	28	41	30

  a Is this data qualitative or quantitative?
     If it is quantitative, is it discrete or continuous?

  b Calculate the median, mode and range of the data. (A tally chart will be helpful.)

  c Display the data and make a statement about your findings.

2   120 people were asked which of three cheeses they liked the most. The results are shown in the pie chart. Calculate how many people preferred each kind of cheese.

3   A die is rolled 20 times, and the scores are as follows:

  5, 6, 2, 1, 5, 4, 6, 3, 2, 6, 6, 1, 4, 2, 5, 6, 1, 2, 4, 3

  Put these results into a frequency table, and find the mean, the mode, the median and the range of the data.

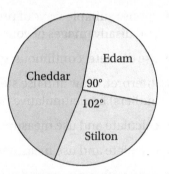

1.1 Measures of central tendency

Types of data

Information obtained from various sources is called **data**.
There are two distinct types of data: qualitative and quantitative.

Qualitative data is extremely varied in nature and includes all data that is not numerical. Qualitative data can often be descriptive data, given as categories – such as hair colour, car type or favourite chocolate bar. It takes no numerical value, although the frequency with which it occurs is numerical. Another name for qualitative data is **categorical data**. Such data can often be represented sensibly by a bar chart, pie chart or pictogram.

Quantitative data, or **numerical data**, is given in numerical form and can be further split into two categories – discrete and continuous. If all possible values that the variable can take can be listed, the data is **discrete**. Examples of discrete data are shoe size, clothes size or the number of marks in a test. **Continuous** data can be shown on a number line, and all points on the line have meaning and are different (for example, someone's height or time to run 100 metres), whereas discrete data can only take a particular selection of values.

In general, discrete data is the result of counting (for example, the number of people in a room), while continuous data is the result of measuring (for example, the combined mass of all the people in the room). However, take care – mass measured to the nearest kilogram is strictly speaking discrete data, since now the data can only take a particular selection of values (we are, in effect, counting whole kilograms). If discrete data is given to a large number of significant figures, you could put it into classes and treat it as if it were continuous data.

Example 1

Are the following sets of data qualitative or quantitative?
If quantitative, is the data discrete or continuous?

A Hair colour of students in a class: {2 black, 7 brown, 6 blonde, …}

B Temperature of water in an experiment: {34.56, 45.61, 47.87, 56.19, … }

C Score shown on two dice in a board game: {2, 6, 7, 7, 8, 9, 12, … }

D Number of spectators at a football match: {12 134, 2586, 6782, 35 765, … }

E Favourite pieces of fruit: {melon, papaya, rambutan, …}

F Times in seconds between 'blips' of a Geiger counter in a physics experiment: {0.23, 1.23, 3.03, 0.21, 4.51, …}

G Scores out of 50 in a maths test: {20, 24, 43, 45, 49, …}

H Size of epithelial cells: {1.20×10^{-5} m, 1.21×10^{-5} m, …}

I Shoe sizes in a UK class: {6, 7, 7, 7, 8, 9, 10, 10, 11, …}

Solution

A is a qualitative data set.

B is a quantitative data set. A temperature reading is a continuous data item, but this is measured to two decimal places, and is strictly discrete. In practice we would treat this as continuous.

C is a quantitative data set. The data is discrete.

D is a quantitative data set. The data is discrete, but in practice we would group this data, and we would in effect treat this data as continuous.

E is a qualitative data set.

F is a quantitative data set. A time reading is a continuous data item, but this is measured to two decimal places, and is strictly discrete. In practice we would treat this as continuous.

G is a quantitative data set. If the score can only be from the set {0, 1, 2, 3, … , 50}, then it is discrete. If the score can take any value between 0 and 50, this starts as continuous data, but since it has been rounded to the nearest mark, it becomes discrete.

H is a quantitative data set. A reading of size is a continuous data item, but if this is measured to three significant figures, it becomes discrete. In practice we would treat this as continuous.

I is a quantitative data set. The data is discrete.

> ❯ Qualitative data is not numerical, but categorical.

> ❯ Quantitative data, or numerical data, can be subdivided into data that is discrete or continuous.

> ❯ Discrete data can only take separate values, such as whole numbers.

> ❯ Continuous data can be shown on a number line, and all points on the line are possible readings for the variable.

Measures of central tendency are often our first tools for comparing and interpreting data. In your previous study you will have encountered three measures of central tendency: the median, the mode and the mean. You will need to be confident in deciding which measure is the most appropriate to use to answer a specific question.

> Measures of central tendency are sometimes referred to as averages or measures of location.

Median

You will recall that when the data is arranged in numerical order, the **median** is the value of data in the middle.

You should use the median for quantitative data, particularly when there are extreme values (values that are far above or below most of the other data) that may skew the outcome.

For example, these data sets with five items each both have median 3.

> 1   2   3   4   20   and   1   2   3   4   5

The extreme value 20 does not affect the value of the median.

Mode

You will recall that the **mode** is the most commonly occurring item of data. It is the item with the highest frequency. There can be more than one mode, if more than one item has the highest frequency, and so the distribution is **bimodal.**

You should use the mode with qualitative data (car models, for example) or with quantitative data (numbers) with a clearly defined mode. The mode is not much use if the distribution is evenly spread, as any conclusions based on mode will not be meaningful.

Mean

You will recall that the **mean** is the sum of all of the items of data divided by the number of items of data.

> When people discuss the average, they are usually referring to the mean.

The formula is normally written as $\bar{x} = \dfrac{\sum x}{n}$, where n is the number of data items.

$\bar{x}$ stands for the mean and is pronounced 'x bar'.

> The Greek letter capital sigma (Σ) means 'the sum of', so $\sum x$ is the sum of all the data.

You should use the mean for quantitative data (numbers). As the mean uses all the data, it gives a true measure (it gives every data item a say), but it can be affected by extreme values.

For example, the data sets with five items introduced earlier have means 6 and 3 respectively.

> 1   2   3   4   20   and   1   2   3   4   5

The difference is entirely due to the value 20.

When a salary increase is being negotiated, the management may have a different opinion to the majority of workers.

The following figures in dollars are the salaries in a small fast-food company in East Timor:

> 3500, 3500, 3500, 3500, 4500, 4500, 4500, 8000, 10 000, 10 000, 10 000, 12 000, 12 000, 18 000, 30 000

Who do you think earns what? The median salary is $8000, the modal salary is $3500 and the mean salary is $9166.67. These figures can be used in a variety of ways, but which is the most appropriate measure? If you were the manager, you might quote the mean of $9166.67, but fewer than half of the employees earn this amount.

Workers leading the pay negotiations who want to criticise the current wage structure may choose to quote the mode ($3500), as this is the lowest average. This would highlight issues in the wage structure.

The mean takes account of the numerical value of every item of data. It is higher due to the effect of the $30 000 salary, which is an extremely large value in comparison to the others. The median is not affected by extreme values.

Advantages and disadvantages of measures of central tendency

The mean can be a good measure to use, since you employ all your data to work it out. It can, however, be affected by extreme values and by distributions of data that are not symmetrical.

The median is not affected by extremes, so it is a good measure to use if you have extreme values in your data, or if you have data that is not symmetrical.

The mode can be used with all types of data, but some data sets can have more than one mode, which is not helpful.

Example 2

Ali rolls a die a number of times, but he does not tell you how many times. You know that his die may be biased towards or away from the value 4 (in other words, the probability of rolling a 4 is unknown) but the other five values are equally likely to be rolled. You know Ali's frequencies for all the values except for 4, and you call this frequency x.

Die score	1	2	3	4	5	6
Frequency	2	3	4	x	2	3

a What values of x make the mode smallest? What values make it largest?

b What values of x make the median smallest? What values make it largest?

c What values of x make the mean smallest? What values make it largest?

d Complete the following table, including the relevant values for x.

	Mode	Median	Mean
Largest possible value			
Smallest possible value			

Solution

a If $x < 4$, the mode is 3. If $x = 4$, the distribution is bimodal. If $x > 4$, the mode is 4. These are the only possible values for the mode.

b If $x < 4$, the median is 3. If $x = 4$, the median is the average of 3 and 4, or 3.5. If $x > 4$, then the median is 4. These are the only possible values for the median.

c The mean is given by:

$$\frac{1 \times 2 + 2 \times 3 + 3 \times 4 + 4 \times x + 5 \times 2 + 6 \times 3}{14 + x}$$

$$= \frac{4x + 48}{14 + x} = \frac{(4x + 56) - 8}{14 + x} = 4 - \frac{8}{14 + x}$$

As x increases, the mean gets larger, but it can never quite reach 4. The smallest value for the mean is when $x = 0$, which gives 3.43 (3 s.f.).

d

	Mode	Median	Mean
Largest possible value	4 ($x \geqslant 5$)	4 (for $x \geqslant 5$)	≈ 4 (for large x)
Smallest possible value	3 (x = 0 to 3)	3 (x = 0 to 3)	3.43 (3 s.f.) ($x = 0$)

Exercise 1.1A

1 State whether the following data is discrete or continuous:

a daily rainfall in Penang

b monthly texts you send on your mobile phone

c the number of burgers sold in a fast food restaurant

d the duration of a marathon

e the ages of the teachers in your school.

2 Classify the following as qualitative or quantitative, discrete or continuous:

a gender   **b** height

c IGCSE grades in maths   **d** examination scores in maths

e waist size   **f** whether people are car owners or not

g weekly self-study time.

3 The number of visits to a library made by 20 children in one year is recorded below.

0, 2, 6, 7, 5, 9, 12, 43, 1, 0, 45, 2, 7, 12, 9, 9, 32, 11, 36, 13

a What is the modal number of visits?

b What is the median number of visits?

c What is the mean number of visits?

d Comment on the best measure of central tendency to use and why.

 Communication    Problem solving    Mathematical Modelling

(PS) **4** Farah records the amount of rainfall, in mm, at her home, each day for a week. The results are:

$$2.8, 5.6, 2.3, 9.4, 0.0, 0.5, 1.8$$

Farah then records the amount of rainfall, x mm, at her house for the following 21 days. Her results are:

$$\Sigma x = 84.6\,\text{mm}$$

a Calculate the mean rainfall over the 28 days.

b Farah realises she has transposed two of her figures. The number 9.4 should be 4.9 and 0.5 should be 5.0. She corrects these figures. What effect will this have on the mean?

(MM) **5** An employee has to pass through 8 sets of traffic lights on her way to work. For 100 days she records how many sets of lights she is stopped at. Here are her results:

Number of times stopped	Number of journeys
0	0
1	3
2	5
3	11
4	21
5	22
6	17
7	14
8	7

Find the median, mode and mean number of times she is stopped at traffic lights and comment on the best measure of central tendency to represent the data.

6 During a biological experiment, 320 fish of various breeds were measured, to the nearest cm, one year after being let into a pond. The distribution of the lengths is shown in the table.

Length, x (cm)	Frequency
$7.5 \leqslant x < 10$	30
$10 \leqslant x < 15$	70
$15 \leqslant x < 20$	100
$20 \leqslant x < 30$	80
$30 \leqslant x < 35$	40

a Calculate an estimate of the mean length of a fish.

b The following year, the class boundaries are found to have increased by a factor of 1.1. What is the effect on the mean length?

c The year after that, the class boundaries are found to have increased by a cm, where $a > 0$. What is the effect on the mean now?

d If the mean increases by the same amount for each of these years, estimate a.

PS 7 The following data was collected but some information was missed out. Copy and complete the table and confirm the estimate of the mean.

Height, h (cm)	Frequency	
$100 < h \leqslant 120$	5	
$120 < h \leqslant 140$	4	
	12	1800
$160 < h \leqslant 180$	13	
$180 < h \leqslant 200$	8	
		6600

Estimate of the mean = 157.14

PS 8 A farmer owns 24 goats, 4 cows and 51 sheep. The mean mass of the goats is 84.1 kg, the mean mass of the cows is 240.3 kg, and the mean mass of the sheep is 99.2 kg. Find the mean mass of all the animals taken together.

C 9 Car owners are given a questionnaire to fill in about their car. It contains the following questions:

a What is the make of the car?

b How many kilometres per litre does it do on average, rounded to the nearest integer?

c What colour is it?

d What is the model of the car?

e What is the fuel tank capacity?

f What is the range of the car on a full tank?

  i Which of the answers would be qualitative, and which would be quantitative?

  ii Of the quantitative answers, which would be discrete and which continuous?

  iii Write one further qualitative question and one further quantitative one.

10 An art gallery does a survey of the works on sale, reviewing the number of completed years that each work has been on their books. (Works that have been on sale for less than 1 year are not included.) The following table is produced.

Number of years	1	2	3	4	5	6	7	8	9	10	11+
Number of artworks	12	16	15	13	8	5	3	2	2	1	0

a Find the median, the mode and the mean for this set of data.

b Find the quartiles and the interquartile range.

c Draw a box-and-whisker plot to show this data.

1.2 Measures of variation

Measures of variation show how spread out, or scattered, data items are. In your previous study you will already have met two measures of variation – the range and the interquartile range.

Range

The simplest measure of variation is the difference between the highest and lowest items of data, known as the **range**. This is straightforward to calculate, but can be highly sensitive to extreme values.

$$\text{Range} = \text{highest value} - \text{lowest value}$$

For example, the data sets with five items introduced earlier have ranges 19 and 4 respectively.

| 1 | 2 | 3 | 4 | 20 | and | 1 | 2 | 3 | 4 | 5 |

The difference once again is due to the large value 20.

Interquartile range

The range is $Q_4 - Q_0$.

LQ   UQ

Q_0 Q_1 Q_2 Q_3 Q_4

Median

If the data set is placed in order and then divided into four groups that are as equal as possible in size, then the three dividing points are the **quartiles**, written as Q_1, Q_2 and Q_3. You have already met the value Q_2 as the **median**. Q_1 is called the **lower quartile** and Q_3 is called the **upper quartile**. Q_0 is the lowest value and Q_4 is the highest value (Q_0 and Q_4 are used less often).

For a small data set, the quartiles are the median of the left-hand half of the data, the median of all the data, and the median of the right-hand half of the data.

For example, here are the quartiles for:

1. a small data set of 10 items

2. a small data set of 11 items

3. a small data set of 12 items.

Q_1 Q_2 Q_3
1   3 ③ 4 ⑥ ⑥ 7 ⑨ 10   11

Where two values are contained in a blue ellipse, it means 'take the average'.

Q_1 Q_2 Q_3
1   3 ③ 4   6 ⑥ 7   9 ⑩ 11   11

Q_1 Q_2 Q_3
1   3 ③ ④ 6 ⑥ ⑦ 9 ⑩ ⑪ 11   12

So for the third set, $Q_1 = 3.5$, $Q_2 = 6.5$, $Q_3 = 10.5$.

The **interquartile range** (or IQR) is defined as $Q_3 - Q_1$ (the range of the middle half of the data). The IQR is unaffected by extreme values, and is therefore generally a more trustworthy measure of variation than the range.

Variance and standard deviation

One limitation with quartiles and the interquartile range is that they do not take all data items into account. Variance and standard deviation use all the data items in determining a measure of variation.

Consider a small set of data: {1, 2, 3, 4, 5}. The mean of this data is 3.

The deviation is the difference between the data item and the mean, usually notated as $x - \bar{x}$.

The set of deviations for this set of data is:

$$1 - 3 = -2, \quad 2 - 3 = -1, \quad 3 - 3 = 0, \quad 4 - 3 = 1, \quad 5 - 3 = 2$$

Adding the deviations gives:

$$-2 + -1 + 0 + 1 + 2 = 0$$

The average deviation from the mean must be 0; it would not be the mean otherwise!

Therefore using sigma notation, $\sum(x - \bar{x}) = 0$, whatever data items you start with. This is no help towards a measure of variation. However, the deviations $x - \bar{x}$ can be combined more sensibly by squaring each deviation (making each positive) and then adding them together. For the small data set above, this gives:

$$(-2)^2 + (-1)^2 + 0^2 + 1^2 + 2^2 = 10$$

Dividing by 5 (the number of data items) to find an average gives 2, which is called the **variance** of the data set. Taking the square root of this (1.414…) gives the **standard deviation** of the data set. Both the variance and the standard deviation are widely used as measures of variation. This chapter focuses on the standard deviation.

The standard deviation of x is denoted by s.d.(x) (or $\sum_x$), while the variance of x is denoted by var(x). The formulae for these are:

$$\text{var}(x) = \frac{\sum(x - \bar{x})^2}{n}, \text{s.d.}(x) = \sqrt{\frac{\sum(x - \bar{x})^2}{n}}$$

since var(x) = (s.d.(x))^2. There is a second formula for both the variance and the standard deviation, which are often easier to work with if you are dealing with raw data:

$$\text{var}(x) = \frac{\sum x^2}{n} - \bar{x}^2, \text{s.d.}(x) = \sqrt{\frac{\sum x^2}{n} - \bar{x}^2}$$

> In general, the IQR is used alongside the median, and the variance and standard deviation are used alongside the mean.

> If the data is measured in a certain unit, the standard deviation will be measured in the same unit, while the variance will be measured in that unit squared.

> There are definitions of the variance and standard deviation that use $n - 1$ instead of n in the denominator. There are good reasons for this, but they are beyond the scope of this course. If n is large, the difference in the definitions is small. Small data sets are useful to demonstrate a concept, but in real-life statistical work, you use large samples that are as representative of the population as possible.

Example 3

Find the standard deviation for the data set {1, 4, 5, 6, 14} using both formulae. Comment on the relative ease of use of each formula.

Solution

Using the first formula:

$$\sum(x-\bar{x})^2 = (1-6)^2 + (4-6)^2 + (5-6)^2 + (6-6)^2 + (14-6)^2$$
$$= (-5)^2 + (-2)^2 + (-1)^2 + 0^2 + 8^2 = 94$$

and so s.d.$(x) = \sqrt{\dfrac{94}{5}} = 4.34\,(3\,\text{s.f.})$

> Most scientific calculators will allow you to input the data item by item and then calculate whatever statistics you require.

Using the second formula:

$$\sum x^2 = 1 + 16 + 25 + 36 + 196 = 274$$

and $\bar{x}^2 = 36$

So

$$\text{s.d.}(x) = \sqrt{\dfrac{274}{5} - 36} = 4.34\ (3\,\text{s.f.})$$

In this case, both methods are easy to use. But when the deviations from the mean are not whole numbers, the second method is more straightforward.

Stop and think   Why must the variance and the standard deviation be positive for any data set?

Standard deviation is especially useful when analysing the position of an item of data in a population (for example, how many standard deviations is it from the mean?). An advantage of standard deviation is that it uses all of the data, while a disadvantage is that it can take longer to calculate than other measures of variation.

Exercise 1.2A

1. An avant-garde piece of music is in 13 sections, and the length of each section in seconds is given below.

   $$24, 36, 16, 48, 37, 35, 27, 45, 11, 43, 50, 44, 9$$

   Find the range and the interquartile range of these times.

2. There were 15 international rugby union matches between six teams in 2014 (everyone played everyone else once). The length of a rugby union game is 80 minutes, but every game runs over. The over-runs in seconds for the 15 games were as follows:

   $$45, 126, 64, 10, 302, 49, 67, 72, 34, 623, 24, 89, 63, 45, 92$$

   Calculate the variance and standard deviation of these times.

PS 3. Give 11 data readings so that their mean is zero and their interquartile range is half the standard deviation.

4 An artist paints tiles. She times herself across one particular day.

She paints 18 tiles in the following times, each measured to the nearest minute.

6, 11, 5, 14, 15, 8, 9, 15, 10, 11, 11, 20, 9, 11, 14, 7, 7, 8

Find the range, the interquartile range and the standard deviation for this set of data.
Which of these measures do you think most accurately describes the variation in the data?

5 Bottles of fruit juice are filled automatically by a machine, but varying amounts of juice are dispensed each time. The contents of a random sample of 12 bottles are measured, with the following results, in cm^3:

330.2   332.0   328.5   335.2   338.7   329.1

331.7   328.5   334.2   329.9   336.4   330.7

Find the mean (to 1 decimal place) and standard deviation (to 3 significant figures) for this data set.

 6 In **Example 3** you found the standard deviation for the small data set {1, 4, 5, 6, 14} using both the first and second formulae.

Since the mean of the data was a whole number, the two methods were equally easy.

Suppose the value 14 is changed to 15, so the mean of the data set is no longer an integer.
Find the standard deviation using both formulae now.

7 A class of 16 boys and 13 girls take a test. The scores for the boys have a mean of 64.4, while the girls have a mean of 68.6. The variance of the scores for the boys is 10.2, while the variance for the scores of the girls is 9.1. Find the mean and variance of the scores for the whole group.

8 Are the following statements true or false?

a The standard deviation of a population is always smaller than the variance.

b The range of a set of data is always greater than or equal to the interquartile range.

c The mean is always greater than the variance.

d The interquartile range is always equal to half the variance.

 9 A statistician is counting the number of sweets in a bag. He takes 30 bags at random to get 30 readings, and for this data set calculates the mean number of sweets in each bag to be 10 and the variance to be 12.

The next day he takes another 30 random readings. By coincidence, he finds that this time the mean is exactly 12, and the variance is exactly 10.

a Work out the mean and variance of the combined set of readings.

b Can you generalise this result for samples of size n, where the mean and variance of the first sample are a and b, and of the second sample are b and a? Show that the variance of the combined set of data is always numerically greater than the mean.

10 You are researching the weekly number of hours worked by the 342 employees in a company.
Discuss the advantages and disadvantages of taking as a measure of spread:

a the range

b the interquartile range

c the variance

d the standard deviation.

11 Jia is playing a video game in which it is possible to score between 0 and 100.

She plays seven games. Her scores on the first six are:

$$34, 54, 24, 37, 39, 42$$

What does she have to score on her final game (which is her best score) if:

a the range of her seven scores is 50

b the IQR of her seven scores is 20

c the variance of her seven scores is 112

d the standard deviation of her seven scores is $10\sqrt{2}$?

1.3 Presenting data

When you are working with raw (unprocessed) information, especially numerical information, there are a number of diagrams that you could use to summarise the data. You should always consider which diagram will convey the meaning of the data most effectively and realistically.

You will remember types of representations including box-and-whisker plots, cumulative frequency diagrams and histograms from your previous study. Now you will need to be able to make judgements about which method of presentation is most appropriate for given data sets.

Stem-and-leaf diagrams

You are given the heights of 28 students (15 boys, 13 girls) in a class, recorded to the nearest cm. The data is as follows:

Boys: 152, 174, 169, 149, 157, 164, 157, 167, 160, 153, 170, 180, 155, 162, 164

Girls: 151, 164, 154, 170, 161, 145, 154, 160, 148, 166, 148, 157, 173

In your previous study you will have covered how to arrange the boys' or the girls' results into a **stem-and-leaf diagram**. The two sets of data can also be displayed side by side on the same stem, as shown below.

Girls' heights			Boys' heights
8 8 5	14	9	
7 4 4 1	15	2 3 5 7 7	
6 4 1 0	16	0 2 4 4 7 9	
3 0	17	0 4	
	18	0	

Key: 1|15|2 = 151 cm for girls and 152 cm for boys

The units digits form the leaves, while the hundreds and tens digits form the stem. You can see at a glance that this particular set of girls are on average shorter than this particular set of boys. A key is always needed for a stem-and-leaf diagram. If the data had been identical but with each item divided by 100, the stem-and-leaf diagram would look the same, but the key would be:

Key: |15|1 = 151 cm

Because the data in a stem-and-leaf diagram is ordered, you can use it to find the quartiles (including the median), the range and the interquartile range for a set of data.

Boys' heights

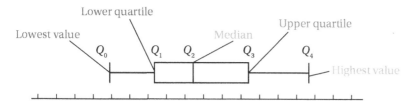

14	9
15	2 3 ⑤ 7 7
16	0 ② 4 4 7 ⑨
17	0 4
18	0

$Q_1 = 155$, Q_2 = median = 162
$Q_3 = 169$, IQR = 14, range = 31

Key: $|16|7 = 167$ cm

> If you turned a stem-and-leaf diagram on its side it would look like a bar chart. The main difference is that stem-and-leaf diagrams preserve the original data.

Box-and-whisker plots

The median and quartiles of a data set can be displayed graphically using a **box-and-whisker plot**, sometimes just referred to as a box plot.

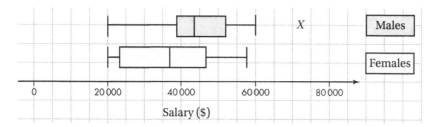

The diagram above shows a standard box-and-whisker plot. There is a horizontal scale. A box is drawn between each of the lower and upper quartiles and a line is drawn in the box showing the position of the median. Whiskers extend from the lowest value to the highest value.

Range = $Q_4 - Q_0$ Interquartile range = $Q_3 - Q_1$

It is useful to be able to compare sets of data using their box-and-whisker plots. The box-and-whisker plots below represent the annual salaries of 30-year-olds in South Korea in 2010.

> **KEY INFORMATION**
> Range = $Q_4 - Q_0$
> Interquartile range = $Q_3 - Q_1$

The ranges of salary are similar, shown by the distance between the whiskers. The males have a smaller interquartile range than females, shown by the size of the boxes, which suggests that the majority of the pay is less spread out for males. The median and quartiles for males are higher than those for females, so on average males earn more than females at age 30. The X on the male diagram indicates a data item which does not seem to fit the trend, an exceptionally high male salary result, called an **outlier**.

There are two commonly used definitions of an outlier.

1. Any value that is more than two standard deviations away from the mean of the data.

2. Any value that is above the upper fence, defined as $Q_3 + 1.5(Q_3 - Q_1)$, or below the lower fence, defined as $Q_1 - 1.5(Q_3 - Q_1)$.

Stop and think You can compare two data sets using a back-to-back stem-and-leaf diagram or two box-and-whisker plots side by side. Which aspects of the data are shown best by each diagram?

Histograms

Histograms are best used for large sets of data when the data has been grouped into classes. They are most commonly used for continuous data. In your previous study you will have encountered histograms with bars of equal width. Now you will learn how to deal with the case where the bars have varying widths, representing unequal class intervals.

The vertical axis of a histogram represents 'frequency density'. This is calculated using the following formula:

$$\text{frequency density} = \frac{\text{frequency}}{\text{class width}}$$

On all histograms, the vertical axis should be labelled 'frequency density' or 'f.d.'

Example 4

You survey 427 professions in Singapore, and record the median salary, rounded to the nearest dollar, for a 30-year-old in each profession. You tabulate your results and draw a histogram.

Salary ($)	Frequency	Class width ($000s)	Frequency density
0–9999	21	10	2.1
10000–19999	129	10	12.9
20000–29999	172	10	17.2
30000–49999	98	20	4.9
50000–99999	7	50	0.14

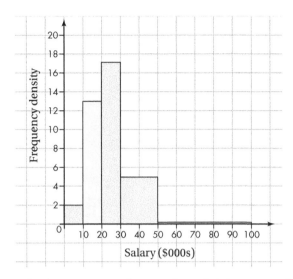

Salary ($000s)

The data is fairly represented, even though it is grouped into intervals with different widths.

Look at the bar shown in blue. The width is 10 and the frequency density is 12.9. So the area of the bar is $10 \times 12.9 = 129$, which equals the frequency.

KEY INFORMATION
For histograms, the frequency is represented by the area of the bar, not by its height.

Drawing histograms

To draw a histogram, you need to calculate the 'class width' and the 'frequency density' for each group. The horizontal axis represents the classes and the vertical axis represents the frequency densities. There should be no gaps between consecutive bars unless the frequency of a particular group is zero. Sometimes, this means that discrete data must be treated as if it were continuous, as shown in **Example 5**.

Example 5

200 people were surveyed to find out how many emails they received on a particular day. The results are shown in the table below.

Number of emails received (n)	Frequency
$1 \leqslant n \leqslant 10$	40
$11 \leqslant n \leqslant 25$	45
$26 \leqslant n \leqslant 50$	60
$51 \leqslant n \leqslant 60$	30
$61 \leqslant n \leqslant 100$	25

a Represent this data on a histogram.

b Use your histogram to estimate the number of people who received less than 40 emails.

Solution

a To ensure that there are no gaps between consecutive bars, the number of emails, which is discrete data, is treated as continuous data. In practice, this means that the class-widths are broadened to close the gaps (as shown in the table below). The calculations for the class widths and frequency densities are as follows:

Number of emails received (n)	Frequency	Class width	Frequency density
$0.5 \leqslant n \leqslant 10.5$	40	10	4
$10.5 < n \leqslant 25.5$	45	15	3
$25.5 < n \leqslant 50.5$	60	25	2.4
$50.5 < n \leqslant 60.5$	30	10	3
$60.5 < n \leqslant 100.5$	25	40	0.625

Plotting the groups of the number of emails on the horizontal axis and the frequency densities on the vertical axis:

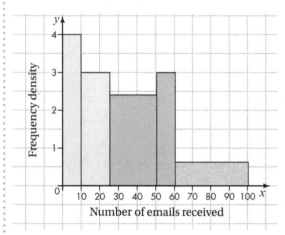

b '40 emails' occurs within the third block on the histogram. To estimate the frequency, you first find out how far into the block the '40 emails' is located. The area of the 3rd block from 25.5 to 40 is $(40 - 25.5) \times 2.4 = 34.8$. You already know that the first 2 blocks represent 85 people in total. Therefore, the number of people who received less than 40 emails is estimated as $85 + 34.8 = 119.8$. Clearly, the number of people must be an integer, so the estimate is rounded to 120 people.

Mathematics in life and work: Group discussion

You are working as a doctor in an area that is suffering from famine. You are studying the effects of malnutrition on children, and whether there is a link between the number of children a mother has and the chance a child has of dying early. You are presented with data about 314 mothers and their children. The results are in the table below.

Number of children	1	2	3	4	5	6	7	8+	Total
Frequency	50	68	112	41	25	15	3	0	314

1   How could you best display this data?

2   How could you challenge the reliability of this data? For example, how were these mothers chosen? What else do you need to consider?

3   How could you use the data that has been collected on child mortality? What diagrams might you use to compare boys' mortality with girls' mortality?

4   What other factors could also affect a child's chance of dying early?

Exercise 1.3A

1   Write down two quantitative variables about your class. Identify each variable as discrete data or continuous data.

2   The lengths of runner beans were measured to the nearest whole cm. Eighty observations of these measurements are given in the table below.

x	3–8	9–13	14–25
Frequency	47	22	11

On a histogram representing this data set, the bar representing the 3–8 class has a width of 2 cm and a height of 4 cm. For the 9–13 class find:

a   the width

b   the height of the bar representing this class.

3   The following table summarises the distances, to the nearest km, that 7359 people travelled to attend a small festival.

Distance (km)	Number of people
40–49	67
50–59	124
60–64	4023
65–69	2981
70–84	89
85–149	75

a   Give a reason to justify the use of a histogram to represent these data.

b   Can you explain why the width of the class interval 40–49 is 10 and not 9?

c   Calculate the frequency densities needed to draw a histogram for these data.

C **4** Here are the number of runs that Jenson and Molly scored during 11 cricket games.

Jenson: 3, 21, 45, 66, 12, 4, 12, 65, 46, 55 and 31

Molly: 34, 57, 12, 98, 17, 22, 17, 43, 23, 76 and 44

a Draw box-and-whisker plots showing their scores.

b Compare their scores, using your diagrams and calculations to help you.

5 The marks of 45 students randomly selected from those who sat a statistics test are displayed below.

36, 39, 39, 40, 41, 42, 42, 43, 44, 45, 46, 46, 46, 48, 50, 52, 53, 53, 54, 54, 55, 55, 56, 57, 57, 59, 60, 60, 60, 60, 61, 63, 64, 64, 64, 65, 65, 66, 67, 68, 69, 71, 72, 73, 73

a What is the modal mark?

b What are the lower quartile, median and upper quartile?

c Represent the data as a box-and-whisker plot.

MM **6** An internet trading company surveys all of the 135 budget printers that it offers on its website. Their prices are given in the table below.

Price of printer ($)	Frequency
$50 < x \leqslant 70$	24
$70 < x \leqslant 100$	31
$100 < x \leqslant 130$	42
$130 < x \leqslant 160$	16
$160 < x \leqslant 200$	18
$200 < x \leqslant 300$	4

Draw a histogram to show this data, and interpret your diagram.

PS **7** This diagram shows the raw test marks of two student groups for the same test.

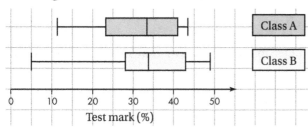

a Use it to determine which of the following statements are true.

i The median is the same for both classes.

ii Q_1 for Class A is 5 marks less than Q_1 for Class B.

iii The interquartile range for Class A is $\frac{2}{3}$ the interquartile range of Class B.

iv $Q_3 - Q_2$ is the same for both classes.

b Write a comparison between the two student groups.

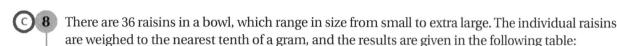
C 8 There are 36 raisins in a bowl, which range in size from small to extra large. The individual raisins are weighed to the nearest tenth of a gram, and the results are given in the following table:

Mass of raisin (g)	Frequency
$0.0 < x \leqslant 0.5$	4
$0.5 < x \leqslant 1.0$	8
$1.0 < x \leqslant 2.0$	14
$2.0 < x \leqslant 3.0$	6
$3.0 < x \leqslant 3.5$	4
3.5–	0

a Draw a histogram to show the data.

b Which class is the modal class (the one with the highest frequency density)?

> Modal class is the class with the highest frequency density.

C 9 A class of 27 students each attempted a mathematical problem without any preparation. The possible scores were between 0 and 50. They then each received tuition on this type of task before tackling a similar problem, scored in the same way. Their before and after scores are shown below. Arrange them into a back-to-back stem-and-leaf diagram and interpret the results.

Before:

10, 46, 20, 25, 35, 6, 17, 40, 23, 32, 18, 12, 34, 17, 29, 30, 25, 5, 23, 43, 23, 18, 35, 16, 21, 23, 14

After:

34, 26, 23, 30, 40, 19, 21, 38, 22, 37, 29, 18, 36, 15, 37, 30, 49, 15, 48, 50, 30, 25, 34, 29, 39, 26, 28

PS 10 One commonly used definition of an outlier involves an upper fence and a lower fence, where:

$$\text{Upper fence} = Q_3 + 1.5(Q_3 - Q_1)$$
$$\text{Lower fence} = Q_1 - 1.5(Q_3 - Q_1)$$

A data point can be regarded as an outlier if it is above the upper fence or below the lower fence.

A large set of data includes the two values 3 and 83. Each value can be categorised as an outlier using the definition above, but only just. Find the lower and upper quartiles for the data.

1.4 Cumulative frequency graphs

Cumulative frequency graphs are used when the data is grouped. It is straightforward to find the median and quartiles using cumulative frequency graphs.

Below is a grouped frequency table showing the time spent queuing for rides at a theme park. It shows, for example, that 41 people queued for between 10 and 20 minutes.

You can work out the cumulative frequencies by adding the frequencies as you go along.

Time, t (minutes)	$0 < t \leqslant 5$	$5 < t \leqslant 10$	$10 < t \leqslant 20$	$20 < t \leqslant 30$	$30 < t \leqslant 60$
Frequency	3	24	41	17	15

Time, t (minutes)	$0 < t \leqslant 5$	$0 < t \leqslant 10$	$0 < t \leqslant 20$	$0 < t \leqslant 30$	$0 < t \leqslant 60$
Cumulative frequency	3	27	68	85	100

Notice that the time intervals also change as the frequencies accumulate.

You can plot these cumulative frequencies on graph paper. Notice that you plot the cumulative frequencies at the upper bound of the interval, that is, at (5, 3), (10, 27), (20, 68), (30, 85) and (60, 100). You should also plot the point (0, 0).

Join the points with a smooth curve to create a cumulative frequency curve, like the one below.

Alternatively, you can join the points with straight line segments to create a cumulative frequency polygon. In practice, they generally give similar answers – if you have plenty of points.

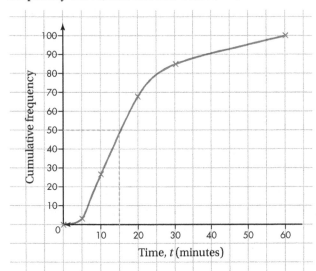

You can find the median by drawing a line horizontally across from 50 (half of 100) on the cumulative frequency axis to the graph, then down to the time axis. In the above diagram you can see that the median is approximately 15 minutes.

Stop and think   Evaluate the statement: 'Cumulative frequency curves are always increasing.'

Estimating percentiles from grouped data

A **percentile** is a value below which a certain percentage of scores fall. For example, if you achieved a score above the 40th percentile, your score would be higher than 40% of the other scores.

The 40th percentile can also be written as P_{40}.

You have already encountered some specific percentiles:

> the 25th percentile, known as the first quartile ($Q_1 = P_{25}$)

> the 50th percentile, the median or second quartile ($Q_2 = P_{50}$)

> the 75th percentile, the third quartile ($Q_3 = P_{75}$).

If data is only available as a grouped frequency distribution, then it is not possible to find the exact values of specific percentiles; you would only be able to estimate them.

For the example above, on the time taken to queue for a ride, to find the two values within which the central 60% of the data lies, you need to find from the graph, the range of values between P_{20} and P_{80}.

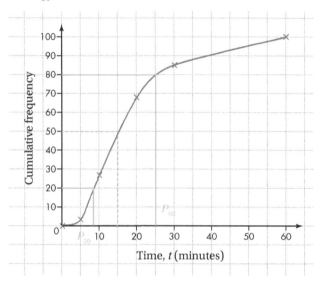

You can see that the 20th percentile is around 8, while the 80th percentile is around 25, so the central 60% of the times lie between 8 and 25 minutes.

This problem was made easier by the fact that the sample was of size 100. Suppose you wish to estimate the 45th percentile of a data set with 31 items. You can calculate that:

$$\frac{45}{100} \times 31 = 13.95$$

so you need to find the 14th data item. Then construct the cumulative frequency curve, and use the horizontal line from 14 on the vertical axis to find P_{45}.

A cumulative frequency curve is a good way to calculate the IQR, which is the range of the central 50% of the data. Use the curve to find P_{25} and P_{75}. The IQR is the difference between these two values.

Example 6

The percentage marks scored in a driving theory test done by 200 members of the public in one day were as follows:

Mark (%)	Frequency
$1 < x \leq 10$	3
$10 < x \leq 20$	11
$20 < x \leq 30$	13
$30 < x \leq 40$	18
$40 < x \leq 50$	26
$50 < x \leq 60$	33
$60 < x \leq 70$	45
$70 < x \leq 80$	34
$80 < x \leq 90$	11
$90 < x \leq 100$	6

Only 45% of the people who did the test that day passed. Estimate the pass mark.

Solution

Mark (%)	Frequency	Cumulative frequency
$1 < x \leq 10$	3	3
$10 < x \leq 20$	11	14
$20 < x \leq 30$	13	27
$30 < x \leq 40$	18	45
$40 < x \leq 50$	26	71
$50 < x \leq 60$	33	104
$60 < x \leq 70$	45	149
$70 < x \leq 80$	34	183
$80 < x \leq 90$	11	194
$90 < x \leq 100$	6	200

45% of people passed, so 55% of people failed. The 55th percentile is needed. This is the 110th data item, which lies in the 61-70 class interval.

The data is continuous – a mark of 60.6% would be rounded to 61%.

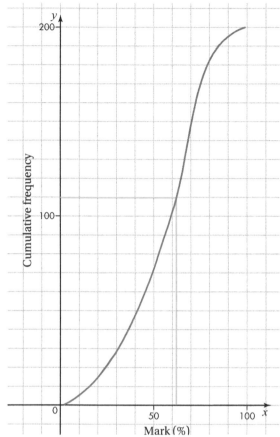

In this example, a frequency curve has been drawn. An alternative method would have been to draw a frequency polygon instead, which involves drawing straight lines between the plotted points.

From the graph, the 55th percentile, P_{55}, is around 63. So, an estimate of the pass mark is 63%.

Mathematics in life and work: Group discussion

You are working as a doctor in India. There has been an outbreak of an infectious disease, and you have been monitoring the situation in one affected area. You have been collecting data on recovery time for people in one village who have now recovered completely from the disease, as shown in the table.

Days to complete recovery from start of infection	0–2	3	4	5	6	7	8	9+
Frequency	0	11	24	56	64	38	20	0

You draw a cumulative frequency graph for this data.

1 What helpful statistics could you calculate from this graph?

2 In particular, what are the median and the interquartile range for this data set?

3 Someone catches the disease and wants to know the chance they will need more than four days to recover. Could you help?

Exercise 1.4A

1 This is a frequency table for the number of people in a household. Estimate the median number of people in the household using a cumulative frequency diagram.

Number in household	Number of households
1	15
2	20
3	22
4	23
5	11
6	4

2 A football coach measured the distance a random sample of 120 11-year-old children could kick a football. The lengths are summarised in the table.

Kick distance, l (m)	Number of children
$5 \leqslant l < 10$	5
$10 \leqslant l < 20$	53
$20 \leqslant l < 30$	29
$30 \leqslant l < 50$	15
$50 \leqslant l < 70$	11
$70 \leqslant l < 100$	7

a Display this data as a cumulative frequency graph.

b Estimate P_{40} for this distribution.

c Estimate the mean of the distribution.

PS **C** **3** The number of aphids on a farmer's strawberry field were counted. The results are presented below.

Number of aphids	Number of strawberry plants
0–19	38
20–29	97
30–39	173
40–49	225
50–69	293
70–99	174

a Construct a cumulative frequency curve and estimate the median.

b Estimate the range within which the central 70% of the results lie. If P_a is the bottom end of this range and P_b is the top end, what are a and b?

c Find an estimate of the mean and compare this with your estimate of the median.

MM **4** The marks of 45 students randomly selected from those who took a statistics test are displayed below.

> 36, 39, 39, 40, 41, 42, 42, 43, 44, 45, 46, 46, 46, 48, 50, 52, 53, 53, 54, 54, 55, 55, 56, 57, 57, 59, 60, 60, 60, 60, 61, 63, 64, 64, 64, 65, 65, 66, 67, 68, 69, 71, 72, 73, 73

   a What is the modal mark?

   b What are the lower, median and upper quartiles?

   c Represent the data as a cumulative frequency diagram.

   d What is $P_{90} - P_{10}$? Why is this a useful measure?

MM **5** A knockout football tournament features 63 games. Any matches that ended in a draw were decided on penalties. The number of goals in these games are given below.

1, 5, 2, 1, 1, 0, 4, 7, 3,

2, 8, 3, 5, 1, 4, 3, 0, 2,

0, 2, 2, 5, 8, 4, 1, 2, 0,

3, 5, 7, 3, 4, 2, 1, 5, 2,

5, 2, 2, 0, 3, 7, 3, 1, 2,

3, 5, 2, 7, 6, 1, 2, 0, 6,

2, 6, 3, 4, 1, 0, 4, 2, 6

Draw a cumulative frequency diagram for this data, and use it to estimate the median.

C **6** Consider this grouped frequency distribution:

Length (cm)	Frequency
$17.5 \leqslant x < 18.0$	15
$18.0 \leqslant x < 18.5$	27
$18.5 \leqslant x < 19.0$	18
$19.0 \leqslant x < 20.0$	12
$20.0 \leqslant x < 25.0$	15
$25.0 \leqslant x < 30.0$	4

   a Identify the upper class values, given that the measurements were recorded to the nearest mm.

   b Construct a cumulative frequency diagram of the data.

   c Using your graph, estimate the median.

   d Compare the median and the mean.

7 This diagram shows the time spent using mobile apps by 200 17-year-olds.

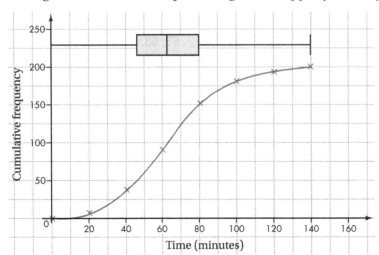

Use the diagram to estimate the following statistics:

a the median time spent using apps   **b** the range of time spent using apps

c $Q_3 - Q_1$ **d** P_{20}

e the number of sessions that lasted longer than 1.6 hours

f compare the box plot and the cumulative frequency diagram shown as different ways to display the same data.

8 A long match is tested by holding the match at an angle and measuring the time it takes to burn out, to the nearest second. The results are given below.

Time (seconds)	26	27	28	29	30	31	32	33+
Frequency	1	3	15	35	23	4	1	0

Draw a cumulative frequency curve for this data, and use it to estimate the width of the interval of times spanning the central 70% of the data.

 9 Which of the following graphs could be parts of cumulative frequency diagrams? Give reasons for your answers.

a
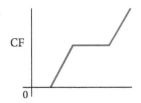

b
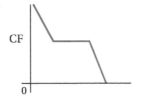

c
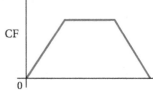

d
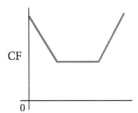

e

f
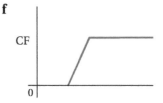

10 A graph has the equation:

$y = 40x + 80$ for $-2 \leqslant x \leqslant 8$,

$y = 400$ for $8 \leqslant x \leqslant 20$, and

$y = 20x$ for $20 \leqslant x \leqslant 30$.

Show that this can represent a complete and viable cumulative frequency curve, where y = cumulative frequency.

Find the median and the interquartile range for this set of data.

1.5 Calculating the mean and standard deviation using frequency tables and summary data

Calculating the mean and standard deviation using frequency tables

To find the mean and standard deviation for data that is in the form of a frequency table, you need to adjust the formulae for $\bar{x}$ and s.d.(x).

Example 7

The table below shows the number of pieces of fruit eaten by students in one day. Find the standard deviation of the number of pieces of fruit eaten.

Fruit, x	Frequency, f
1	2
2	12
3	45
4	114
5	41
6	16
Total	**230**

Solution

Fruit, x	Frequency, f	fx	x^2	fx^2
1	2	2	1	2
2	12	24	4	48
3	45	135	9	405
4	114	456	16	1824
5	41	205	25	1025
6	16	96	36	576
Total	**230**	**918**	**91**	**3880**

First multiply the number of fruit by its corresponding frequency to give fx (column 3), then sum that column to find the total number of items of fruit, $\sum fx$.

Next, square each value to give x^2 (column 4), then sum that column to find $\sum x^2$.

Finally, multiply x^2 by the frequency to give fx^2 (column 5), then sum that column to find $\sum fx^2$.

For a frequency table like this, the formulae become:

$$\text{s.d.}(x) = \sqrt{\frac{\sum f(x-\bar{x})^2}{\sum f}} \text{ or s.d.}(x) = \sqrt{\frac{\sum fx^2}{\sum f} - \bar{x}^2}$$

$\sum f$, the sum of the frequencies, is equal to n.

For our example, the mean is $\frac{918}{230}$

$$\frac{3880}{230} - \left(\frac{918}{230}\right)^2 = 0.939\,05$$

Taking the square root, $\sqrt{0.939\,05} = 0.969$, and so

$\text{s.d.}(x) = 0.969$ (3 s.f.)

> For grouped frequency distributions you use the mid-point of the group as an estimate for x, just as you did for calculating an estimate of the mean.

> In squaring the mean, the error is also squared, and sometimes this can make a dramatic difference. Use as accurate a value of the mean as possible when calculating the standard deviation.

Mathematics in life and work: Group discussion

You are a doctor monitoring the heart rates of the patients at your practice. You measure the resting heart rates of patients, and record your data in the table below.

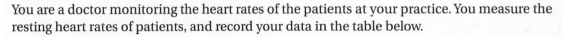

Heart rate at rest (beats/min)	$20 < x \leqslant 30$	$30 < x \leqslant 40$	$40 < x \leqslant 50$	$50 < x \leqslant 60$	$60 < x \leqslant 70$	$70 < x \leqslant 80$	$80 < x \leqslant 90$	$90 < x \leqslant 100$	$100 < x \leqslant 110$	$110 < x \leqslant 120$	$120 < x \leqslant 130$
Frequency	1	0	2	24	56	64	48	32	11	5	2

You have to decide whether any of your patients have a heart rate that can be regarded as exceptionally high or low (an outlier).

1 What precautions would you need to take in collecting this data?

2 What are the best estimates you can make of the mean, standard deviation and interquartile range of this data?

3 You work with a definition of an outlier as any result that is more than two standard deviations from the mean. Will any of your patients be classified as outliers?

4 You work with an alternative definition for an outlier as any value $1.5 \times \text{IQR}$ outside the interval (Q_1, Q_3). Are there any outliers in your data using this definition?

Using summary data

Sometimes you will be presented not with the raw data to analyse, but with statistics that summarise the raw data. For example, imagine you are a doctor given the times in days (*t*) that a cast is in place for 1000 patients with a minor leg fracture. Instead of being given 1000 individual scores, you are told that:

$$\sum t = 50\,902, \sum t^2 = 2\,656\,452$$

The mean of the data is $\dfrac{\sum t}{n} = \dfrac{50\,902}{1000} = 50.902$, while the standard deviation is:

$$\sqrt{\dfrac{\sum t^2}{n} - \bar{t}^2} = \sqrt{\dfrac{2\,656\,452}{1000} - 50.902^2} = 8.09 \ (3 \text{ s.f.})$$

So you can say that a patient with a minor leg fracture is likely to need a cast for around 51 days with a standard deviation of around 8 days.

Using coding

You collect a set of values for a variable *x*. You could now transform your data values into values of a new variable *y* by saying $y = x + a$, where *a* is some constant. This can make the calculation of $\bar{x}$ and s.d.(*x*) simpler. This is called **coding** the data.

For example, you wish to find the mean and standard deviation of the following set of data:

Blood volume, *x*	Frequency
1247.3	4
1248.3	12
1249.3	15
1250.3	12
1251.3	8

The raw data does not look easy to work with, but you can code the data as follows:

$$y = x - 1247.3$$

Construct a new table, subtracting 1247.3 from all the *x* values:

Coded blood volume, *y*	Frequency
0	4
1	12
2	15
3	12
4	8

Calculating the mean and standard deviation for this second table is straightforward. You find:

$$\bar{y} = \frac{\sum fy}{\sum f} = \frac{4 \times 0 + 12 \times 1 + 15 \times 2 + 12 \times 3 + 8 \times 4}{51} = 2.16 \text{ (3 s.f.)}$$

$$\text{s.d.}(y) = \sqrt{\frac{\sum y^2}{n} - \bar{y}^2} = \sqrt{\frac{308}{51} - \left(\frac{110}{51}\right)^2} = 1.18 \text{ (3 s.f.)}$$

Note that $\frac{110}{51}$ is used for the mean (an exact value) rather than 2.16 (a rounded value).

So how do you now convert back to x, the variable you collected data for at the start?

Stop and think Suppose you have a set of data, and you shift every data item to make it 15 units smaller. What is the effect on the mean? What is the effect on the standard deviation?

When you coded the data above, you said that $y = x - 1247.3$.

You can now state the following:

$$\bar{y} = \bar{x} - 1247.3 \qquad \text{s.d.}(y) = \text{s.d.}(x)$$

So

$$\bar{x} = 1247.3 + 2.16 = 1249.46 \qquad \text{s.d.}(x) = 1.18$$

Normally you would not leave the mean to 6 significant figures, but the spread is so low here that this is a reasonable thing to do.

So in general, if $y = x + a$, then $\bar{y} = \bar{x} + a$, and s.d.$(y) = $ s.d.(x)

Using coding with summary data

Sometimes you may need to work with summary statistics that have been coded. Suppose the severity of a patient rash (x) is recorded on a continuous scale from 20 to 30. You, the doctor, are given data for 100 patients in the form:

$$\sum(x - 20) = 398, \qquad \sum(x - 20)^2 = 1891$$

What is the mean $(\bar{x})$, and what is the standard deviation (s.d.(x)), of the data?

What should your coding be? A natural choice is to let $y = x - 20$.

That means from the summary statistic that $\bar{y} = \dfrac{398}{100} = 3.98$

You also have that:

$$\text{s.d.}(y) = \sqrt{\frac{1891}{100} - 3.98^2} = 1.75 \text{ (3 s.f.)}$$

So converting back from y to x, $\bar{x} = 20 + \bar{y} = 23.98$, and s.d.$(x) = $ s.d.$(y) = 1.75$

Example 8

The speeds (in km/hour) of 25 bicycles passing a particular point are recorded. The information is summarised into two pieces of information:

$$\sum(x-10) = -5, \quad \sum(x-10)^2 = 2803$$

a Find the mean, variance and standard deviation of these speeds.

b Find the value of $\sum x^2$.

Solution

a If $y = x - 10$, then $\bar{y} = -\dfrac{5}{25} = -0.2$ so $\bar{x} = \bar{y} + 10 = 9.8$ km/h.

$$\text{s.d.}(y) = \sqrt{\frac{2803}{25} - (-0.2)^2} = 10.6 \ (3 \text{ s.f.}) = \text{s.d.}(x)$$

Therefore $\text{var}(x) = \dfrac{2803}{25} - (-0.2)^2 = 112.08 = 112.1 \ (4 \text{ s.f.})$

b Rearranging the formula for variance:

$$\sum x^2 = 25(\text{var}(x) + \bar{x}^2) = 25(112.08 + 9.8^2) = 5203$$

Example 9

Simona measured the mass in kg of 40 random sacks of sand. The results are summarised below.

$$\sum(x-60) = -814 \qquad \sum(x-60)^2 = 22\,125$$

Find the mean and standard deviation for the mass of the sacks of sand.

Solution

Let $y = x - 60$

$$\bar{y} = \frac{-814}{40} = -20.35$$

$$\text{s.d.}(y) = \sqrt{\frac{22\,125}{40} - (-20.35)^2} = 11.79$$

So $\bar{x} = \bar{y} + 60 = -20.35 + 60 = 39.65$ kg and s.d.$(y) = $ s.d.$(x) = 11.79$ kg

Exercise 1.5A

1 Calculate the mean and standard deviation of the following data sets without a calculator. Check your results using a calculator.

  a 2, 4, 6, 8, 10, 12, 14

  b 50, 60, 70, 80, 90

  c 12, 15, 18, 16, 7, 9, 14

2 For each of the following sets of data, find the mean and standard deviation.

  a 2, 2, 4, 4, 4, 5, 6, 6, 8, 9

  b 13.1, 20.4, 17.4, 16.5, 21.0, 14.8, 12.6

Ⓒ **3** The table gives the shoe sizes of a class of British students.

Size	3	4	5	6	7	8	9
Frequency	1	0	8	14	6	2	1

  a Calculate the mean and standard deviation of the set of data.

  b What can you say about the mean size compared with the median and modal sizes?

4 You are given that for a particular data set, $\Sigma x = 27$, $\Sigma x^2 = 245$, $n = 13$.

  Find the mean and standard deviation of the data.

5 The length of time, t minutes, taken for a bus journey is recorded on 15 days and summarised by:

$$\Sigma x = 102, \qquad\qquad \Sigma x^2 = 1181$$

  Find the mean, variance and standard deviation of the times taken for this bus journey.

(PS) **6** For a set of 20 data items, $\Sigma(x - 10) = 12$ and $\Sigma(x - 10)^2 = 144$.

  a Find the mean and standard deviation.

  b Find $\Sigma(x - 8)$ and $\Sigma(x - 8)^2$.

  c For a second data set of 30 items, $\Sigma(y - a) = 48$ and $\Sigma(y - a)^2 = 314$. If the mean for each data set is the same, find the variance of the y-values.

(MM) **7** Mrs Moat has a choice of two routes to work, a town route and a country route. She times her journeys along each route on five random occasions, and the times in minutes are given below.

Town route	15	16	20	28	21
Country route	19	21	20	22	18

  a Calculate the mean and standard deviation for each route.

  b Which route would you recommend? Give a reason.

  c Mrs Moat travels again along the town route in p minutes, and the variance for the six town times now has the value 20. To the nearest integer, what are the possible values for p?

PS 8 A bag of nails states that it contains 60 nails. A sample of 40 bags is chosen at random, and the following results for x, the number of nails in a bag, are found:

$$\sum x = 2641, \sum x^2 = 175\,042$$

a find the mean and the standard deviation of x

b if, for some data set of size 40, $\sum x = 2641$, what is the smallest possible value of $\sum x^2$?

MM 9 A farmer measures the mass of each apple in a batch of 200 apples. He puts the results into the following table:

Mass, m (g)	$50 < x \leqslant 60$	$60 < x \leqslant 65$	$65 < x \leqslant 70$	$70 < x \leqslant 75$	$75 < x \leqslant 80$	$80 < x \leqslant 90$	$90 < x \leqslant 120$
Frequency	14	28	31	46	45	24	12

a Estimate the mean and standard deviation for the mass of the apples.

b One extra apple is then added to the batch; it has a mass equal to the mean of the sample of 200. Explain what effect adding the apple will have on the mean and standard deviation of the new batch of 201 apples.

10 A drug is administered to 155 patients on one day in a hospital. The doses are classified into six volumes as follows:

Volume, v (mm^3)	Frequency
$122.5 < x \leqslant 132.5$	19
$132.5 < x \leqslant 142.5$	24
$142.5 < x \leqslant 152.5$	35
$152.5 < x \leqslant 162.5$	41
$162.5 < x \leqslant 172.5$	27
$172.5 < x \leqslant 182.5$	9

Find estimates for $\sum (v - 127.5)$ and $\sum (v - 127.5)^2$ and hence estimate the mean and variance of v.

11 Alexandra counted the number of cars passing her house between 9 am and 9:15 am each day for 20 consecutive days. The results are summarised below.

$$\sum (x - 25) = 124 \qquad \sum (x - 25)^2 = 3531$$

Find the mean and standard deviation for the number of cars passing Alexandra's house between 9 am and 9:15 am each day.

12 Kamal counted the number of sweets in a random selection of 50 bags of sweets. The results are summarised below.

$$\sum (x - 70) = -315 \qquad \sum (x - 70)^2 = 2567$$

Find the mean and standard deviation for the number of sweets in each bag.

SUMMARY OF KEY POINTS

> Qualitative data are not numerical, but categorical.

> Quantitative data, or numerical data, can be subdivided into discrete or continuous.

> Discrete data may only take separate values, such as whole numbers.

> Continuous data may take any measured value, usually within a range.

> Measures of location (averages) include the median, the mode and the mean.

> The median is the middle value when the data is placed in numerical order.

> The mode is the most common or most frequent item of data.

> The mean is found by adding the data values together and dividing by the number of values: $\bar{x} = \dfrac{\sum x}{n}$.

> Measures of variation include the range, the interquartile range, the variance and the standard deviation.

> Standard deviation = s.d.$(x) = \sqrt{\sum \dfrac{(x-\bar{x})^2}{n}} = \sqrt{\dfrac{\sum x^2 - n\bar{x}^2}{n}} = \sqrt{\dfrac{\sum x^2}{n} - \bar{x}^2}$.

> Variance = var(x) = (standard deviation)^2.

> Box-and-whisker plots display the quartiles together with the minimum and maximum points.

> Histograms can be used with either equal widths or unequal widths to display continuous data. It is the area of a bar, and not its height, that represents frequency.

> Cumulative frequency curves can be used to find the quartiles and percentiles, and the interquartile range.

> Summary statistics may be given, and the mean and standard deviation can be calculated from these.

EXAM-STYLE QUESTIONS

MM **1** Here is a data set showing the heights of employees in a company rounded to the nearest cm.

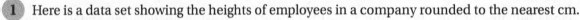

1.45, 1.48, 1.46, 1.52, 1.46, 1.61, 1.60, 1.51, 1.55, 1.56,

1.61, 1.64, 1.53, 1.51, 1.48. 1.70, 1.70, 1.62, 1.45, 1.50

  a Draw a stem-and-leaf diagram to show this data.

  b Draw a box-and-whisker plot to provide an alternative representation.

C **2** It is reported in the news that teenagers use social media for a long time each day. A random sample of 11 students were interviewed and asked how long they spent using social media in an average week.

The total durations, y, in minutes, for the 11 students were:

7, 98, 121, 132, 151, 187, 204, 255, 260, 277, 357

  a find the median and quartiles for this data

  b show that there are no outliers, where an outlier is defined as any result more than $1.5 \times$ IQR outside the range (Q_1, Q_3).

C **3** A class of 15 boys and 15 girls was surveyed to see how highly they scored on a particular mathematical game. The results were as follows:

Girls: 143, 179, 167, 152, 173, 156, 149, 166, 134, 148, 155, 142, 166, 172, 147

Boys: 153, 169, 142, 173, 165, 184, 152, 143, 184, 152, 139, 156, 176, 162, 142

a draw a back-to-back stem-and-leaf diagram with boys on the left-hand side and girls on the right-hand side to show this data set

b find the interquartile range of the girls' scores

c a new boy arrives in the class and when his score is added to the data, the mean score for the boys is 160. What did the new boy score?

  4 A city in the UK has a mean yearly rainfall in 2015 of 100 cm with standard deviation 7 cm.

What rainfall would you expect in order to make it an exceptional year?

(An 'exceptional year' is one in which the rainfall is more than 2 standard deviations away from the mean.)

5 The amount of profit in one year on $50 000 given by six investments, A to F, of differing risk is given in the table below.

Destination	A	B	C	D	E	F
Number of investors	136	278	154	299	113	89
P ($)	−1564	2098	2592	−988	3200	4805

a Work out the mean profit for an investor.

b Work out the standard deviation of the profit for an investor across all six investments.

6 Mr Hardy selects a random sample of 40 students and records, to the nearest hour, the time they spent gaming in a particular week.

Hours	1–5	6–10	11–15	16–20	21–30	31–49
Frequency	3	7	11	14	4	1
Mid-point		8	13	18		40

a Find the mid-points of the 1–5 hour and 21–30 hour groups.

On a histogram representing this data set, the 6–10 group is represented by a bar of width 2 cm and height 5 cm.

b Find the width and height of the 31–49 group.

c Estimate the mean.

7 Here is a table for the variable t, which represents the time taken, in minutes, by a group of people to run 3 km. (Here the dash means 'up to but not including'.)

a Copy and complete the frequency table for t, and draw a cumulative frequency curve to represent the data.

t (minutes)	5–10	10–14	14–18	18–25	25–40
Frequency		15	22		18
Frequency density	2			3	

b Estimate from your curve the number of people who took longer than 20 minutes to run 3 km.

c Find an estimate of the mean time taken.

d Find an estimate for the standard deviation of t.

e Find the median, the quartiles and the 60th percentile for t.

C **8** The table below summarises the lengths of songs in seconds of the most recent 246 number 1 hit singles in the US charts.

Length of song, t (sec)	$130 < t \leq 150$	$150 < t \leq 160$	$160 < t \leq 170$	$170 < t \leq 180$	$180 < t \leq 190$	$190 < t \leq 220$
Frequency	11	24	67	85	42	17

a In which time interval does the median lie? Which time interval contains the upper quartile?

b On graph paper, draw a histogram to represent the data.

c Calculate an estimate of the mean length of a number 1 hit single in the US.

9 A survey of 100 households gave the following results for their monthly shopping bills, y.

Monthly shopping bill, y ($)	Mid-point	Frequency, f
$0 \leq y < 200$	100	6
$200 \leq y < 240$	220	20
$240 \leq y < 280$	260	30
$280 \leq y < 350$	315	24
$350 \leq y < 500$	425	12
$500 \leq y < 800$	650	8

A histogram was drawn and the class $200 \leq y < 240$ was represented by a rectangle of width 2 cm and height 7 cm.

a Calculate the width and the height of the rectangle representing the class $280 \leq y < 350$.

b Estimate the median monthly shopping bill to the nearest dollar using a cumulative frequency diagram.

c Estimate the mean and the standard deviation of the monthly shopping bills for these data.

(c) 10 A group of students carry out an investigation on the lengths of children's feet before they have their teenage growth spurt. The students measured the foot lengths of a random sample of 100 11-year-old children. The lengths are shown in the table.

Foot length, l (cm)	Number of children
$12 \leqslant l < 17$	2
$17 \leqslant l < 19$	14
$19 \leqslant l < 21$	38
$21 \leqslant l < 23$	34
$23 \leqslant l < 25$	12

a Using a cumulative frequency curve, find an estimate of the median length.

b Using a calculator, estimate the mean and standard deviation.

c What type of average would you use to best describe this data?

(c) 11 The amounts of money, x dollars, that a sample of 30 cinema-goers spent on the cinema in the year 2017 is summarised by:

$$\Sigma(x - 160) = 146, \qquad \Sigma(x - 160)^2 = 1024$$

a find the mean and the standard deviation of x directly from the figures 160, 146, 1024 and 30 using addition, subtraction, division and squaring

b find Σx and Σx^2

c find the mean and standard deviation of x from your values for Σx and Σx^2.

(c) 12 Danielle and Hannah decide to weigh cookies served in the canteen.

Here is a grouped frequency table showing the mass in grams of each cookie.

Mass, m (grams)	$30 \leqslant m < 40$	$40 \leqslant m < 50$	$50 \leqslant m < 60$	$60 \leqslant m < 80$	$80 \leqslant m < 120$
Frequency	13	37	56	8	6

a Draw a cumulative frequency graph and use it to estimate:

   i the median mass of a cookie

   ii the lower and upper quartiles of the masses of the cookies.

b Construct a frequency density table and use it to draw a histogram displaying the masses.

c Hannah has heard the rough rule that the standard deviation is usually about three-quarters times the interquartile range. Find the standard deviation for the data above and check if this rule works roughly here.

13 In a powerlifting competition, a weightlifter's best attempts at each category are combined to give their final score. The total of the combined weights, w, in kilograms of 17 powerlifters is 4930 kg and the standard deviation is 10.5 kg.

    a A judge becomes concerned that a weightlifter is attempting to lift too much if they try to lift two standard deviations more than the mean. What is the maximum weight a powerlifter could try to lift before a judge becomes concerned?

Another powerlifter joins the competition and lifts a combined weight of 292 kg.

    b Find the new mean.

14 A group of students are asked to do a blindfolded reaction test. When they hear a beep, they have to click a button which records the speed of their reaction. A random sample of 104 17-year-olds are asked to take part and the results are recorded to the nearest millisecond. The results are summarised in this table.

Time (milliseconds)	Mid-point	Frequency
0–9	4.5	6
10–19	14.5	14
20–29	24.5	34
30–39	34.5	27
40–49	44.5	19
50–99	74.5	4

In a histogram, the group '10–19 milliseconds' is represented by a rectangle 2.5 cm wide and 4 cm high.

    a Calculate the width of the rectangle representing the group 50–99 milliseconds.

    b Calculate the height of the rectangle representing the group 20–29 milliseconds.

    c Using a cumulative frequency diagram, estimate the median and interquartile range.

    d The estimate for the mean for the data is 30.2. Would you use the mean or the median when reporting your findings?

 15 An agriculturalist is studying the mass, m kg, of courgette plants. The data from a random sample of 70 courgette plants is summarised below.

Yield, m (kg)	Frequency, f
$0 \leqslant m < 5$	11
$5 \leqslant m < 10$	29
$10 \leqslant m < 15$	18
$15 \leqslant m < 25$	8
$25 \leqslant m < 35$	4

(You can use $\sum fx = 750$ and $\sum fx^2 = 11\,312.5$)

A histogram has been drawn to represent this data.

The bar representing the mass $5 \leqslant m < 10$ has a width of 1.5 cm and a height of 6 cm.

a Calculate the width and the height of the bar representing the mass $25 \leqslant m < 35$.

b Estimate the mean and the standard deviation of the mass of the courgette plants.

C 16 A survey is carried out in 2016 in a region of India into the age of drivers involved in car crashes severe enough to see the car written off. The results are as follows:

Age	18–27	28–37	38–47	48–57	58–69	70–89
Frequency	105	64	42	30	21	10

a find (to the nearest month) the mean and standard deviation of these ages

b what is the modal class

c find the median and the interquartile range by using a cumulative frequency curve

d would you consider the mean, the mode or the median to be the best measure of average here? Give your reasons.

PS 17 A special pack of cards consists of cards showing the integers from 1 to 25. A sample of 11 cards is taken at random as follows, where $a > 10$:

$$5, 2, 9, 10, 3, 7, 1, 8, 4, 6, a$$

The value a is such that the interquartile range for the sample is almost equal to the standard deviation; in fact, for no other card would the standard deviation be closer to the interquartile range. Find a.

18 A cycling club measures the time taken by each of its 125 members to complete a 20 km course. The times (in seconds) can be summarised as follows:

$$\sum t = 310\,467, \; \sum t^2 = 773\,071\,329$$

a Find the mean and the standard deviation for this set of data, in each case to the nearest second.

b The club statistician complains that these figures are large, and the chance of a mistake is great. He suggests using a coding. Find $\sum (t - 2500)$, $\sum (t - 2500)^2$ for these figures.

C 19 A cricket writer claims that the standard of fitness amongst first-class cricketers in Australia is higher than that in the English first-class players. As a part of a charity fun day, 27 randomly chosen first-class cricketers in Australia and 27 in England are timed carrying out 50 sit-ups. Their times to the nearest tenth of a second are as follows:

Australia: 48.3, 49.1, 53.5, 50.6, 51.3, 49.9, 51.8, 52.6, 52.1, 50.1, 49.8, 49.0, 52.8, 51.0, 51.4, 49.2, 48.5, 50.9, 50.4, 51.8, 53.4, 50.6, 48.8, 52.1, 53.2, 51.5, 50.7

England: 51.4, 50.8, 49.2, 49.6, 53.6, 50.4, 51.9, 50.8, 50.1, 48.0, 52.3, 51.6, 49.3, 49.9, 52.0, 50.1, 48.4, 51.9, 51.2, 49.7, 48.8, 51.6, 50.7, 52.4, 48.1, 49.3, 50.8

a Display this data as a back-to-back stem-and-leaf diagram.

b Display this data as two box-and-whisker plots over the same scale.

c What conclusions can you draw from your diagrams? Which method of display do you feel is most effective? Give reasons for your answers.

20 The following facts are true for two different sets of data:

    a their highest values are the same at 30

    b their ranges are the same at 22

    c their interquartile ranges are the same at 12

    d the median of the first set equals the lower quartile for the second set

    e the upper quartile of the first set equals the median for the second set

    f for the first set of data, $Q_3 - Q_2 = 2(Q_2 - Q_1)$

    g for the second set of data, the lower quartile is 14.

Draw box-and-whisker plots representing each set of data over the same scale, and comment on the two distributions.

Mathematics in life and work

You are a doctor working at NICE (National Institute for Health and Care Excellence), the body that decides on which drugs the National Health Service in the UK is allowed to use to fight illness.

A drug has to prove that it gives a good return for its cost.

> For each drug, the average extension of life, L, for a patient is found in years.

> Then a measure is made of the quality of life, Q, for a patient on that drug over that time: 1 is the quality of life that a completely well person would enjoy, while 0 represents a terrible quality of life.

> The measure $L \times Q$ gives the quality years achieved by being on the drug.

> The cost $\$C$ per person per drug treatment is also measured.

> The measure of cost-effectiveness is CE = $\dfrac{C}{L \times Q}$, which is the cost per quality year.

   NICE cannot recommend the continuation of drugs that have a CE score of more than 7.5.

For a particular type of cancer, six different drugs, A to F, are being considered over the course of a trial. The patients column gives the number of patients (203 in total) who are using each drug at a particular hospital.

Drug	Patients	L	Q	$L \times Q$	C ($1000)	CE	Continue?
A	15	1	0.9		10		
B	31	2	0.7		12		
C	64	3	0.7		15		
D	42	4	0.6		8		
E	34	5	0.4		17		
F	17	6	0.3		3		

1 Complete the missing cells in the table.

2 Calculate the mean value and standard deviation for C over all patients.

3 Find the median and the quartiles for the values of L over all patients.

4 For which drug is CE largest?

5 Which drugs appear to you to be the most cost-effective? Which ones would the team at NICE want to discontinue?

2 PROBABILITY, PERMUTATIONS AND COMBINATIONS

Mathematics in life and work

Uncertainty surrounds many real-life situations in which you have to make a decision. You use the concept of probability every day. You might ask, 'What will the temperature be today?' or 'What is the probability of all the traffic lights being green on my journey home?' or 'What is the likelihood of passing your driving theory test first time?' These situations demand decisions and you can use the mathematics of probability to help you make them.

Closely linked to the idea of probability is the notion of a permutation or a combination. For example, if you have five books to place on a shelf, how many different arrangements are there? If you have a bag containing five different chocolates and you choose two at random from the bag, how many different ways are there to do this? This chapter will help you to answer these questions.

Many careers involve working with probabilities. For example:

- If you were working as a doctor and studying the spread of a disease, you might need to know the likelihood that one person will pass it to another.

- If you were a sports journalist, you might need to work out the probability that a darts player will hit a treble 20.

- If you were an actuary you would need to calculate risk, for example, in business, healthcare and insurance.

This chapter includes some of the problems you might encounter if you were an actuary.

LEARNING OBJECTIVES

You will learn how to:

- solve problems involving permutations and combinations of a set of objects
- model situations involving probability and explain any assumptions made
- evaluate probabilities in simple cases
- use sample spaces in simple cases
- add and multiply probabilities in appropriate cases
- use both Venn diagrams and tree diagrams to calculate probabilities
- show that events are independent or mutually exclusive
- use conditional probability in simple cases
- use the conditional probability formula $P(A \mid B) = \dfrac{P(A \cap B)}{P(B)}$.

LANGUAGE OF MATHEMATICS

Key words and phrases you will meet in this chapter:

- combination, complement, conditional probability, dependent events, event, factorial, independent events, mutually exclusive events, permutation, probability, sample space diagram

PREREQUISITE KNOWLEDGE

You should already know how to:

› complete tables and grids to show outcomes and probabilities

› interpret P(A) as the probability of event A

› interpret P(A') as the probability of event A not occurring

› use the complement of an event, A′, where P(A') = 1 − P(A)

› complete a tree diagram to show outcomes and probabilities

› interpret and use a Venn diagram.

You should be able to complete the following questions correctly:

1 A gym club has 120 members and 89 of the members train for boxing, 54 of the members train for kickboxing, while 30 of the members train for both.

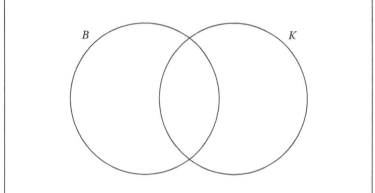

  a Use this information to complete the Venn diagram by inserting these frequencies.

  B represents those members who box.

  K represents those members who kickbox.

  b How many members train for neither boxing nor kickboxing?

2 A box contains 15 pieces of fruit where five are apples, six are oranges and four are pears. If one piece of fruit is selected from the box at random, find the probability that the fruit is:

  a an apple

  b a pear

  c an apple or an orange.

3 The Venn diagram shown on the right contains probabilities. Use it to calculate:

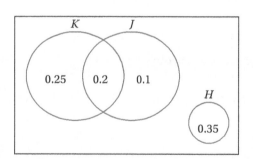

  a P(K and J)　　　　**b** P(K and H)

  c P(H')　　　　　　　**d** P(K' and J' and H')

  e P(K or H).

4 A bag contains ten counters. Six are blue and four are yellow. One counter is selected at random, then replaced. A second counter is then selected.

  a Draw a tree diagram to represent this information.

  b Find the probability that they are both yellow.

  c Find the probability that one is blue and the other is yellow.

2.1 Permutations and combinations

Suppose you write down the digits 1, 2, 3, 4 and 5.

Question A: How many different five-digit numbers can you make from these digits if you are not allowed to repeat a digit?

Consider the first digit – there are five options. Then, for the second, you have a choice of four digits (since you are not allowed repeat digits). So you have $5 \times 4 = 20$ ways of choosing the first two digits. Using this logic three more times, there are $5 \times 4 \times 3 \times 2 \times 1 = 120$ different numbers you can make.

The number $n \times (n-1) \times \ldots \times 2 \times 1$ (for a positive whole number n) is called **n factorial**, or $n!$.

This function grows very quickly as n increases. If you try to calculate $100!$ on a calculator, it will almost certainly give you an error message.

Question B: How many two-digit numbers can you make from the five numbers digits, with no repeat digits?

This is similar to the question above – there are five ways for you to pick the first digit, and four to pick the second, so the answer is 20.

Question C: How many different pairs of digits can you select from the five numbers, with no repeat digits?

You have the possibilities {1, 2}, {1, 3}, {1, 4}, {1, 5}, {2, 3}, {2, 4}, {2, 5}, {3, 4}, {3, 5} and {4, 5} – that is, 10 in total.

A **permutation** has the notation ^nP_r (the notation nPr is sometimes seen) and describes the number of ways of choosing r things from n things if order matters.

A **combination** has the notation ^nC_r and describes the number of ways of choosing r things from n things if order does not matter.

> **Stop and think** Which will be bigger, ^nC_r or ^nP_r? What will ^nP_n be if n is any positive whole number?

Question A involved calculating $^5P_5 = 5!$

Question B involved calculating $^5P_2 = 5 \times 4 = \dfrac{5 \times 4 \times 3 \times 2 \times 1}{3 \times 2 \times 1} = \dfrac{5!}{3!}$

Question C gave the same answer as **Question B**, except each pair was counted twice, so you needed to divide by 2:

$$^5C_2 = \frac{5!}{3!} \times \frac{1}{2} = \frac{5!}{3!2!}$$

These are the formulae for ^nP_r and ^nC_r:

$$^nP_r = \frac{n!}{(n-r)!} \qquad ^nC_r = \frac{n!}{r!(n-r)!}$$

KEY INFORMATION

In general, the number of ways of arranging a list of n different objects in a line without repetition is $n!$.

Make sure you know where the $n!$ button is on your calculator. The number $0!$ is defined as 1.

The order in which you choose the digits does not matter, so the two cases {2, 1} and {1, 2} are only counted once.

An alternative notation for ^nC_r is $\begin{pmatrix} n \\ r \end{pmatrix}$.

(nCr is also sometimes seen.)

KEY INFORMATION

^nP_r describes the number of ways of choosing r things from n things if order matters.

^nC_r describes the number of ways of choosing r things from n things if order does not matter.

Make sure you know which buttons on your calculator you need, to calculate ^nP_r and ^nC_r for any n and r.

Stop and think

Can you see why these formulae are true?

^nP_r is $n \times (n-1) \times (n-2)... \times (n-r+1)$

^nC_r is $n \times (n-1) \times (n-2)... \times (n-r+1)$, but divided by r.

If you need to work out ^nP_r and ^nC_r by hand, you can use these formulae. This may involve a lot of cancelling, for example:

$$^8P_5 = \frac{8!}{5!} = \frac{8 \times 7 \times 6 \times 5 \times 4 \times 3 \times 2 \times 1}{5 \times 4 \times 3 \times 2 \times 1} = 8 \times 7 \times 6 = 336$$

$$^8C_5 = \frac{8!}{5!3!} = \frac{8 \times 7 \times 6 \times 5 \times 4 \times 3 \times 2 \times 1}{5 \times 4 \times 3 \times 2 \times 1 \times 3 \times 2 \times 1} = 8 \times 7 = 56$$

Stop and think

Why must ^nP_r and ^nC_r always be whole numbers?

Can you show that for all n and r,

$^nP_n = n!$, $^nP_0 = 1$, $^nC_n = {}^nC_0 = 1$, $^nC_{n-1} = {}^nC_1 = n$, and $^nC_r = {}^nC_{n-r}$?

The triangle in blue on the left is known as Pascal's Triangle; you add two neighbouring numbers to get the one below.

Pascal's Triangle is in fact built from the values for ^nC_r, which are given in red on the right. Drawing out the first few rows of Pascal's Triangle can be a useful way to find ^nC_r if n is small.

$$^0C_0$$
$$^1C_0 \quad ^1C_1$$
$$^2C_0 \quad ^2C_1 \quad ^2C_2$$
$$^3C_0 \quad ^3C_1 \quad ^3C_2 \quad ^3C_3$$
$$^4C_0 \quad ^4C_1 \quad ^4C_2 \quad ^4C_3 \quad ^4C_4$$
$$^5C_0 \quad ^5C_1 \quad ^5C_2 \quad ^5C_3 \quad ^5C_4 \quad ^5C_5$$
...

Example 1

How many ways are there of arranging the letters of these words in a line?

a FACETIOUS

b NEEDLESS

An arrangement of letters in a line is called a string.

Solution

a Since all the letters in this word are different, the answer is simply $9! = 362\,880$.

b First, suppose that all the letters are different. You can do this by labelling the three Es and two Ss:

$$NE_1E_2DLE_3S_1S_2,$$

which gives $8!$ arrangements.

But within those $8!$ arrangements, there are $3!$ ways to mix up the three Es, and $2!$ ways to mix up the two Ss. That means you need to divide $8!$ by both $3!$ and $2!$, so the number of distinct arrangements is $\frac{8!}{2!3!} = 3360$.

For small strings of letters, an alternative method would be to write down or group the different combinations. This can become increasingly time consuming for strings of more than a few letters.

Example 2

Seven members of a family line up to have a photo taken. Jessie and Tiffany have had an argument, so they cannot stand next to each other. How many arrangements are there for the line?

Solution

With no restrictions, there are 7! possible arrangements. If you treat JT as a single letter, then there are 6! arrangements containing JT, while if you treat TJ as a single letter, there are a further 6! arrangements. So there are $2 \times 6!$ disallowed arrangements, and the number of possible arrangements is:

$$7! - (2 \times 6!) = 3600$$

Example 3

The letters in the word DISAPPOINTED are arranged so all the vowels are together. How many ways are there to do this?

Solution

The vowels are I_1, A, O, I_2, E, so there are $\frac{5!}{2!} = 60$ ways to arrange these. The consonants are D_1, S, P_1, P_2, N, T, D_2. Treating the group of five vowels as an extra letter, there are $\frac{8!}{2!2!} = 10\,080$ allowable ways to arrange the letters. Now you need to multiply by 60 to take account of the possible vowel arrangements, which is $604\,800$ ways.

Example 4

A doctor has a choice of seven small but different bandages, six medium but different bandages, and five large but different bandages. She chooses five bandages, and her choice includes at least one of each size. How many different selections could she have made?

Solution

Her choice must be either a 1–1–3 choice or a 1–2–2 choice.

The number of 1–1–3 choices is:

$$^7C_1 \times {}^6C_1 \times {}^5C_3 + {}^6C_1 \times {}^5C_1 \times {}^7C_3 + {}^7C_1 \times {}^5C_1 \times {}^6C_3 = 2170$$

The number of 1–2–2 choices is:

$$^7C_1 \times {}^6C_2 \times {}^5C_2 + {}^6C_1 \times {}^5C_2 \times {}^7C_2 + {}^5C_1 \times {}^7C_2 \times {}^6C_2 = 3885$$

So the total number of possible choices that the doctor could have made for the bandages is $2170 + 3885 = 6055$.

Example 5

A group of eight friends are the only people watching a film in a cinema. They sit in two rows of four, one behind the other where each row has two ends.

Screen

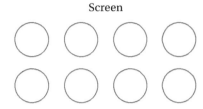

a If anyone can sit anywhere, how many different arrangements of the friends are there?

b If Mark and Sarah must both sit on the end of a row, how many possible seating arrangements are there now?

Solution

a $8! = 40\,320$ ways

b There are four positions for Mark, each of which leaves three positions for Sarah, and each of these 12 possibilities gives rise to 6! ways to arrange the others.

So altogether there are $12 \times 6! = 8640$ possible arrangements.

Example 6

A committee of five people is to be selected from a group of seven men (including Ian and Oscar) and six women (including Vanessa and Natalie). However, the committee must not contain both Ian and Vanessa, and additionally it must not contain both Oscar and Natalie. How many possible committees are there?

Solution

If Ian and Oscar are in the committee, there are $\binom{9}{3}$ possibilities. If Ian but not Oscar is in the committee, or Oscar but not Ian is in the committee, there are $\binom{10}{4}$ possibilities. If neither Ian nor Oscar are in the committee, there are $\binom{11}{5}$ possibilities.

So in total there are $\binom{9}{3} + 2\binom{10}{4} + \binom{11}{5} = 966$ possibilities.

Example 7

A team of acrobats form a human pyramid on stage, as shown below.

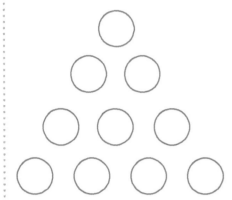

The team contains four large acrobats and six small ones. The large acrobats have to form the bottom row.

a How many possible pyramids could the audience see?

A large acrobat leaves the team to be replaced by a small acrobat. Now the bottom row of the pyramid must consist of the three large acrobats with a small acrobat as one of the central pair.

b How many possible pyramids could the audience see now?

Solution

a There are 4! ways to organise the bottom row, and 6! ways to form the top three rows, so there are
4! × 6! = 17 280 ways.

b There are seven ways to pick the small acrobat for the bottom row, and they can go in one of two positions, so there are 2 × 7 × 3! = 84 ways to organise the bottom row. Once this is done there are 6! = 720 ways to organise the top three rows, so there are 60 480 ways to organise the pyramid altogether.

Exercise 2.1A

1 **a** How many ways can the letters in the word COUNTERS be arranged in a line?

b How many three-letter strings can be made from the letters in COUNTERS if no repeats are allowed?

 2 On each card of a pack of 26 cards is written a different letter of the alphabet. How many arrangements are there for the pack if the first five cards must be the vowels A, E, I, O and U in some order? Give your answer to three significant figures in standard form.

 **C** Communication   **MM** Mathematical modelling   **PS** Problem solving

3 Rohit puts eight books onto his bookshelf. Six are novels (N_1, N_2, N_3, N_4, N_5 and N_6) and two are collections of short stories (S_1 and S_2).

   a How many ways are there of arranging the eight books on his shelf?

   b How many ways are there of arranging the eight books on his shelf if the two books of short stories are not next to each other?

   c If he has to pick four books, at least one of which is a novel and at least one of which is a collection of short stories, in how many ways can he do this?

4 In a tennis competition, there are 16 players left, of whom five are French, four are Dutch, three are American and four are Japanese. Five of these players are chosen for an interview so that at least one player from each nationality is chosen. How many possible choices for the interview group are there?

5 A philosophy student has 12 books with either Plato or Socrates in the title: 8 with Plato, and 4 with Socrates. She arranges these 12 books at random on a shelf.

Find the probability that:

   a all the Plato books are together, and all the Socrates books are together

   b all the Plato books are together.

6 **a** How many different ways are there of arranging the letters in the word ROCOCO?

   b How many different ways are there of arranging the letters in the word ROCOCO if the R must be on one end?

   c How many different ways are there of arranging the letters in the word ROCOCO if a C has to be at each end?

7 **a** Assuming that a repeated letter cannot be distinguished from another appearance of that letter, find the number of distinct arrangements that can be made using all eight letters of the word ELLIPSES.

   b Find the number of these arrangements in which the two Ls do not appear together.

8 The four kings and the four queens from a pack of cards are arranged in a row of eight.

Find the number of ways that:

   a the king of hearts and the queen of hearts are next to each other

   b the king of spades and the king of clubs are not next to each other

   c The king of hearts and the queen of hearts are next to each other AND the king of spades and the king of clubs are also next to each other.

9 A small group of mahjong players with m men and n women decide to form a club. They decide further to choose a committee of four people. They find that the number of ways of choosing two men and two women from their group is $0.9 \times$ the number of ways of choosing one man and three women. What are the smallest values that m and n could be?

(PS) **10** Nine friends are travelling in a van with nine seats, arranged in three rows, as shown below.

 Driver

a If Peter or Alice or Mark must drive, how many arrangements are there?

b If Peter must drive and Mark must sit in the second row and Alice in the third row, how many possible arrangements are there?

c If Peter must drive and Alice and Mark must sit on the end of a row, how many possible arrangements are there?

2.2 Evaluating probabilities

In mathematics, **probability** is the likelihood of an **event** happening (or not happening). An event could be anything from 'obtaining a head when flipping a coin' to 'it will rain next Friday'. A trial, or experiment, has various outcomes, and events are one or more of the outcomes. If A is an event, the number $P(A)$ is short for 'the probability of A happening', and $P(A)$ always lies between 0 and 1 inclusive. If $P(A) = 1$, A is certain to happen, while if $P(A) = 0$ you can be certain A does not happen.

The event 'A does *not* happen' is called the **complement** of A and is written as A'.

Now A must either happen or not happen, so $P(A) + P(A') = 1$.

You take an unbiased die with six faces. The outcomes of rolling the die are 1 to 6.

All the possible outcomes are equally likely here, which means you can use this formula:

$$\text{Probability of an event} = \frac{\text{number of successful outcomes}}{\text{total number of equally likely outcomes}}$$

For example, P(even number) $= \frac{3}{6} = \frac{1}{2}$, while

P(number is 5 or less) $= \frac{5}{6}$.

If a trial consists of two separate activities, a **sample space diagram** is a good way to show your results. For example, if two dice are rolled simultaneously and their scores added, the sample space diagram would look like this:

> **KEY INFORMATION**
>
> The probability of an event A, $P(A)$, is a number such that $0 \leqslant P(A) \leqslant 1$.
>
> If $P(A) = 0$, A does not happen, while if $P(A) = 1$, A is certain to happen.

> **KEY INFORMATION**
>
> The complement of an event A is A', the event A does not happen, and $P(A') = 1 - P(A)$.

> The probability of any given face being on the top when the die is rolled is $\frac{1}{6}$.

> **KEY INFORMATION**
>
> $P(A) = \dfrac{\text{number of outcomes where } A \text{ happens}}{\text{total number of outcomes}}$
>
> $P(A)$ is a number where $0 \leqslant P(A) \leqslant 1$.
>
> $P(A) = 0$ means A does not happen, while $P(A) = 1$ means A is certain to happen.
>
> A must either happen or not happen, so $P(A) + P(A') = 1$.

Die 2	6	7	8	9	10	11	12
	5	6	7	8	9	10	11
	4	5	6	7	8	9	10
	3	4	5	6	7	8	9
	2	3	4	5	6	7	8
	1	2	3	4	5	6	7
	+	1	2	3	4	5	6
		Die 1					

Note that every white cell is an equally likely outcome here. It is easy to see the most common occurring value for the sum is 7. The number 7 appears 6 times out of a total of 36 outcomes, so the probability of rolling a scoring 7 is $P(7) = \frac{6}{36} = \frac{1}{6}$.

Example 8

A game involves flipping a coin and picking a card at random, from four cards numbered 1 to 4.

a Draw a sample space diagram to show all the possible outcomes.

b Find the probability of flipping a head and picking a card with a square number on it.

Solution

a First you need to draw your sample space diagram, thinking about the labels for your headings. Then list all the possible outcomes of the combined events.

Tails	(T, 1)	(T, 2)	(T, 3)	(T, 4)
Heads	(H, 1)	(H, 2)	(H, 3)	(H, 4)
	1	2	3	4

Note that all the outcomes in the table are equally likely.

b Using the sample space diagram, identify the successful outcomes.

Tails	(T, 1)	(T, 2)	(T, 3)	(T ,4)
Heads	**(H, 1)**	(H, 2)	(H, 3)	**(H, 4)**
	1	2	3	4

An alternative method of solving part **b** would be to use a tree diagram.

$$P(\text{heads and square number}) = \frac{2}{8} = \frac{1}{4}$$

Example 9

Liam and Samir are playing a game that involves choosing one of three dice coloured orange, black or white. Liam chooses first, and Samir knows what Liam chooses. Then each player rolls his die, and the one showing the higher number is the winner.

Each die has different numbers on each face:

Orange	5, 7, 8, 9, 10, 18
Black	2, 3, 4, 15, 16, 17
White	1, 6, 11, 12, 13, 14

The total on each die is 57, so the game would appear fair. Do you agree?

> A fair game is one where the probabilities of each player winning are equal.

Solution

In order to model this situation, add in the assumption that the dice are selected randomly to play against each other. The individual dice are unbiased, but is this enough to make the game fair?

To determine if this game is fair, the coloured dice can be played against each other and the winning die can be recorded in a sample space diagram. Then you will see whether any of the dice has an advantage.

> A *fair* die is one where every face is equally likely to be thrown. Such a die is also called *unbiased*. If the faces are not equally likely to be thrown, the die is *biased*.

First match: Orange versus black

The winning die is shown by the initial in the sample space diagram.

> There is a branch of mathematics called Game theory, which is beyond the scope of this course. It offers a range of alternative approaches for analysing this type of situation.

```
           Orange die
        | 5  7  8  9  10 18
      2 | O  O  O  O  O  O
      3 | O  O  O  O  O  O
Black 4 | O  O  O  O  O  O
die  15 | B  B  B  B  B  O
     16 | B  B  B  B  B  O
     17 | B  B  B  B  B  O
```

Probability of orange die winning = $\frac{21}{36}$

Probability of black die winning = $\frac{15}{36}$

The first match results show that the orange die has a higher probability of winning against the black die.

Second match: Orange versus white

		Orange die				
		5	7	8	9	10	18

White die		5	7	8	9	10	18
	1	O	O	O	O	O	O
	6	W	O	O	O	O	O
	11	W	W	W	W	W	O
	12	W	W	W	W	W	O
	13	W	W	W	W	W	O
	14	W	W	W	W	W	O

Probability of orange die winning = $\frac{15}{36}$

Probability of white die winning = $\frac{21}{36}$

The second match results show that the white die has a higher probability of winning against the orange die.

Third match: White versus black

Black die		White die 1	6	11	12	13	14
	2	B	W	W	W	W	W
	3	B	W	W	W	W	W
	4	B	W	W	W	W	W
	15	B	B	B	B	B	B
	16	B	B	B	B	B	B
	17	B	B	B	B	B	B

Probability of white die winning = $\frac{15}{36}$

Probability of black die winning = $\frac{21}{36}$

The third match results show that the black die has a higher probability of winning against the white die.

So if he knows which die Liam has chosen, Samir can always pick a die that gives him an advantage over Liam. The game is biased.

To make the game fair, Liam and Samir could each select their die at random.

You can often use permutations and combinations to calculate probabilities, as shown in **Example 10**.

Example 10

The letters in the word MAKES are arranged in a line at random. What is the probability that E and M are together?

Solution

The total number of arrangements of the letters M, A, K, E and S in a line are 5!.

Treat EM as a single letter. The number of arrangements containing EM is 4!. Similarly, there are 4! arrangements containing ME. So:

P(arrangement has E and M together) = $\frac{2 \times 4!}{5!} = \frac{2}{5}$

Example 11

There are 14 boys and 16 girls in a class. A committee of four students is chosen at random. Find the probability that the committee contains no boys.

Solution

The number of possible selections for the committee is

$$\binom{30}{4} = {}^{30}C_4 = 27\,405.$$

The number of possible choices of four girls is $\binom{16}{4} = {}^{16}C_4 = 1820.$

So the probability that the committee contains no boys is

$$\frac{1820}{27405} = \frac{52}{783} = 0.0664 \text{ (3 s.f.)}.$$

Mathematics in life and work: Group discussion

You are working as an actuary and you have an approximate model for the weather in April on a mountain in Indonesia. You categorise each day's weather into sun, cloud, rain or snow. You believe that on average in April, there are 8 days of sun, 13 days of cloud, 7 days of rain and 2 days of snow. You base this statement on years of records. Your approximate model also states that the weather on a particular day is independent of the weather that comes before it.

1   How many different arrangements of weather are possible across the 30 days in April? Give this to 3 s.f. in standard form.

2   You say that if the 7 days of rain occur together, there is a significant risk of flooding in the towns below. What is the probability (given your model) that the 7 days of rain in April occur together? In how many years (to 1 s.f.) would you expect one such event to occur?

3   How realistic is your claim that the weather on a particular day is independent of the weather on previous days?

4   Does your answer to Question 3 increase or decrease your estimate of the risk of flooding?

Exercise 2.2A

1   An international school has four houses: Rio, Qom, Aba and Nis. Of the 200 students in Year 9, 40 are in Rio, 62 in Qom and 52 in Aba.

   a   What is the probability that a student chosen at random is in Nis?

   b   What is the probability that a student chosen at random is in neither Qom nor Aba?

2   Tickets numbered 1 to 100 are placed in a hat and a single ticket is chosen at random.

   a   What is the probability of selecting a ticket with an even number?

   b   What is the probability of selecting a ticket with a number that is not a square number?

   c   What is the probability of selecting a ticket with a number containing the digit 5?

3 A bag contains 20 balls of which x are black, $2x$ are white and 8 are red. A ball is selected at random from the bag. What is the probability that it is not black?

 4 Aimee has a collection of eight hats, of which three are blue and identical. The others are yellow, green, red, black and purple. She arranges the hats at random on a shelf.

    a What is the probability that all the blue hats are together?

    b What is the probability that the yellow hat and the green hat are not together?

    c What is the probability that the hats on both ends of the line are blue?

5 A park has three kinds of trees, with 10 blackwoods, 13 lychees, and 7 meranti. A selection of four trees is made at random.

    a What is the probability that the selection is made up of only blackwoods?

    b What is the probability that the selection contains no blackwoods?

    c What is the probability that the selection includes at least one of each type of tree?

 6 Alan is an 18-year-old man who moves to Australia to work. Unemployment in Australia is at 13.2% for 16-to-24 year olds, and there are 1.3 million men aged 16–24 living in Australia at that time. What would you estimate is the probability that Alan will get a job?

    Critically analyse your answer, stating any other assumptions you would make and the effect they would have on the probability.

 7 A coin is tossed and a card is chosen at random from a normal pack and its suit is noted.

    a Draw up a sample space for the outcomes.

    b What is the probability you get a head and a red suit?

    The coin is tossed a second time.

    c Draw up a new sample space diagram to show all the possible outcomes now.

    d What is the probability you get a head and a tail (in some order) and a spade?

8 From a football team of 11 players (a goalkeeper, five defenders and five attackers) a team of five players is chosen for a five-a-side competition.

    What is the probability that the team of five contains the goalkeeper, two defenders and two attackers?

 9 An association of golfers is run by a council of 24 members made up of 10 men and 14 women. From the council, a subcommittee of four people is chosen at random, comprising a chair, a secretary and two treasurers.

    a How many different subcommittees can be chosen?

    b What is the probability that the subcommittee contains three women and one man, with a woman as chair and the man as secretary?

    c What is the probability that both treasurers are men?

10 If *A* is an event, then *A'* is the event 'A does not happen'.

A card is picked from a shuffled pack.

a If *H* is the event 'you pick a heart', what is P(*H'*)?

b If *T* is the event 'you pick a 10', what is P((*H'* or *T*)')?

c If *B* is the event 'you pick the 10 of hearts', what is P((*B* or (*T* or *H*))')?

2.3 Venn diagrams and probabilities

Venn diagrams are helpful in discussing and calculating probabilities. The *A* circle contains all events where *A* happens, and the *B* circle contains all events where *B* happens.

The shaded area in this diagram contains the events where *A* and *B* both happen. This area can be described as $A \cap B$, or *A intersection B*, or *A and B*.

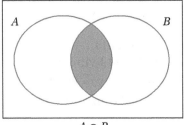

$A \cap B$

> **KEY INFORMATION**
>
> In probability, the event *A* or *B* generally means the event *A* or *B* or both.

The shaded area in the diagram below can be described as $A \cup B$, or *A union B*, or *A or B*. This area contains the events where either *A* or *B* happens, or both happen.

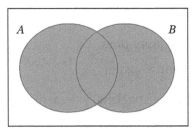

$A \cup B$

Stop and think

Describe the shaded area in the diagram on the left. What about the diagram on the right?

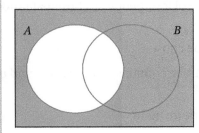 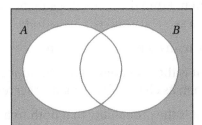

Example 12

In a small college with 100 students, maths and economics are the two most popular subjects.

55 students are studying maths, while 27 students are studying economics.

17 students are studying maths and economics.

Find the probability that a student chosen at random is studying maths or economics or both.

Subject	Number of students	Probability
Maths	55	0.55
Economics	27	0.27
Both	17	0.17

KEY INFORMATION

If A and B are two events then:

› $A \cap B$ represents the event 'both A and B occur'.

› $A \cup B$ represents the event 'either A or B (or both) occur'.

$P(A \cap B)$ and $P(A \cup B)$ give the probabilities of these events.

Solution

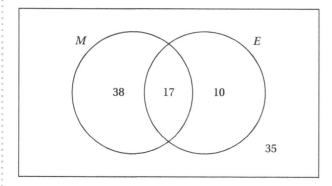

$P(M$ or $E) = 0.38 + 0.17 + 0.10 = 0.65$

An alternative method would be to use the complement. So $P(M$ or $E) = 1 - 0.35 = 0.65$

Sometimes a Venn diagram will contain probabilities that add up to 1, rather than frequencies.

Example 13

A and B are two events and $P(A) = 0.65$, $P(B) = 0.75$ and $P(A \cup B) = 0.85$.

Find:

a $P(A \cap B)$

b $P(A')$

c $P(A' \cup B)$

d $P(A' \cap B)$

Solution

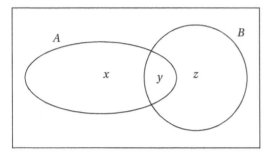

This is a Venn diagram showing probabilities.

You know $x + y = 0.65$, $y + z = 0.75$, and $x + y + z = 0.85$.

Subtracting the first equation from the third equation gives $z = 0.2$.

So $x = 0.1$, $y = 0.55$, $w = 0.15$

a $\mathrm{P}(A \cap B) = y = 0.55$

b $\mathrm{P}(A') = z + w = 0.35$

c $\mathrm{P}(A' \cup B) = w + y + z = 0.9$

d $\mathrm{P}(A' \cap B) = z = 0.2$ •------

> By shading the appropriate parts of the Venn diagram, you will often find alternative ways of describing the same probability, which can lead to alternative methods. For example, in part **d**, $\mathrm{P}(A' \cap B)$ is the same as $\mathrm{P}(B) - \mathrm{P}(A \cap B)$.

Example 14

If a fair die is thrown, what is the probability that it shows:

a that event A is a prime number

b that event B is a number less than 3

c that event A or B is a prime number or a number less than 3?

Solution

a Three of the six numbers on a die are prime: 2, 3 and 5.

So $\mathrm{P}(A) = \frac{3}{6} = \frac{1}{2}$

b Two of the six numbers on a die are less than 3: 1 and 2.

So $\mathrm{P}(B) = \frac{2}{6} = \frac{1}{3}$

c Four numbers on a die are prime or less than 3, or both: 1, 2, 3 and 5.

So $\mathrm{P}(A \text{ or } B) = \frac{4}{6} = \frac{2}{3}$ •------

> $\mathrm{P}(A \text{ and } B) = \frac{1}{6}$, as only 2 is prime and less than 3.

Exercise 2.3A

1 For events A and B, $P(A) = 0.5$, $P(B) = 0.3$ and $P(A$ and $B) = 0.2$.

   a What is $P(A$ or $B)$?

   b What is $P(A'$ and $B)$?

2 A and B are two events such that: $P(A) = 0.3$, $P(B) = 0.5$ and $P(A$ and $B) = 0.15$. Find:

   a $P(A')$

   b $P(B')$

   c $P(A$ or $B')$.

(PS) 3 K and S are two events and $P(K) = P(S) = 3P(K \cap S)$ and $P(K \cup S) = 0.75$. Find:

   a $P(K \cap S)$ **b** $P(K)$ **c** $P(S')$

   d $P(K' \cap S')$ **e** $P(K \cap S')$.

4 You are told that $P(A) = 0.5$, $P(B) = 0.6$ and $P(A' \cap B') = 0.3$. Find:

   a $P(A \cap B)$

   b $P(A \cap B')$

   c $P(B \cap A')$.

(MM) 5 A club of pianists contains 156 members. Among these members, 46 also play the violin and 78 also play the guitar. 10 members play both the guitar and the violin.

A member is picked at random. Find the probability that she or he plays the piano but neither the violin nor the guitar.

(C)(6) In the UK, 40% of households have a juicer and 50% have a laptop, while 25% of households have both. Calculate the probability that a household chosen at random has either a juicer or a laptop but not both.

(C)(7) A survey of 100 gardeners asks which of three garden implements (spade, fork or hoe) they have used in the last week. There are 10 who have used the hoe only, 24 who have used the fork only, and 14 who have used the spade only. There are 7 gardeners who have used all three implements. The numbers of those using exactly two of the implements are all equal. Nine gardeners have used none of these implements.

One gardener is picked at random to receive a prize. Find the probability that this gardener used a spade that week.

(PS)(8) In this Venn diagram, A, B and C are events and x and y are probabilities. You are told that $P(B) = 0.3$. Find $P(C)$.

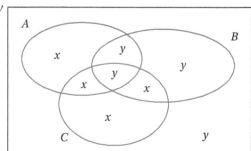

C 9 In a swimming club of 50 young people, 15 do crawl and backstroke, 24 do crawl and butterfly, while 19 do butterfly and backstroke. 30 do crawl, 30 do backstroke, and 33 do butterfly. 7 members do exactly one stroke from crawl, backstroke and butterfly.

A member is picked at random. Find:

a the probability that the member does all three strokes

b the probability that the member does none of the three strokes.

PS 10 A and B are two events such that $P(A' \cap B')$, $P(A \cap B')$, $P(A' \cap B)$ and $P(A \cap B)$ are four consecutive terms from a geometric progression where the common ratio is $r = \frac{1}{2}$. Find $P(A)$ and $P(B)$.

C 11 When Ahmed goes for a walk on a cool day, there is probability of $\frac{2}{3}$ that he wears a hat and a probability of $\frac{1}{2}$ that he wears a scarf. The probability that he wears neither a hat nor a scarf is $\frac{1}{4}$. Some of this information is shown in the table below.

	Wears a scarf	Does not wear a scarf	
Wears a hat			$\frac{2}{3}$
Does not wear a hat		$\frac{1}{4}$	
			1

a Copy and complete the table.

b Given that he does not wear a hat, find the probability that he wears a scarf.

2.4 Mutually exclusive, independent and dependent events

Mutually exclusive events (or just 'exclusive events') are events that cannot both occur at once. For example, a coin toss can have the outcome heads or tails, but not both. A die can land on a 6 or a 1, but not both at once. If A and B are mutually exclusive events, the Venn diagram will look like this:

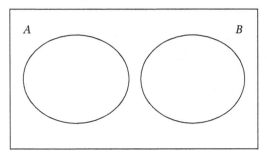

If A and B are mutually exclusive, then:

» $P(A \cap B) = 0$

» $P(A \cup B) = P(A) + P(B)$.

Independent events do not affect each other. This means that the probability of one event occurring, such as rolling a 6 on a die, would not affect the probability of the other event occurring, such as flipping a head on a coin.

Dependent events are events that influence each other.

Suppose you are picking two socks out of your drawer on a dark morning. If you have two red socks, two blue socks and two white socks in the drawer:

› the probability that the first sock you pick is white is $\frac{2}{6}$

› if you do pick out a white sock first, the probability that the next sock is also white is now only $\frac{1}{5}$

› if you don't pick a white sock first, the probability that the second sock is white is now $\frac{2}{5}$.

The two events 'first sock is white' and 'second sock is white' are dependent. The probability of the second event depends on what happens first.

If A and B are independent, then $P(A \cap B) = P(A) \times P(B)$.

We can also say that if $P(A \cap B) = P(A) \times P(B)$, then A and B are independent.

So if you wonder whether two events A and B are independent or not, the best way to tackle this is usually to find $P(A)$ and $P(B)$, and check whether or not $P(A \cap B) = P(A) \times P(B)$.

KEY INFORMATION

If A and B are mutually exclusive they cannot occur together, which means $P(A \cap B) = 0$, and

$P(A \cup B) = P(A) + P(B)$

If A and B are independent, then $P(A \cap B) = P(A) \times P(B)$.

Example 15

In a questionnaire, event Y was 'this participant is in the age group 11–16', while event Z was 'this participant is in the age group 21–25'. Y and Z are mutually exclusive, and for the population surveyed, $P(Y) = 0.3$ and $P(Z) = 0.5$. Find:

a $P(Y \cup Z)$ **b** $P(Y \cap Z')$ **c** $P(Y' \cap Z')$.

Solution

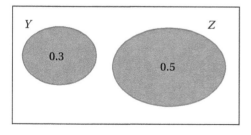

a $P(Y \cup Z) = P(Y) + P(Z) = 0.3 + 0.5$
$\qquad\qquad = 0.8$

b $P(Y \cap Z') = P(Y)$
$\qquad\qquad = 0.3$

c $P(Y' \cap Z') = 1 - P(Y \cup Z) = 1 - 0.8$
$\qquad\qquad = 0.2$

Example 16

A card is picked at random from a pack of 52 cards. What is the probability that a heart or a spade is picked? Are the events 'heart' and 'spade' independent?

Solution

The events 'you pick a heart' and 'you pick a spade' are mutually exclusive, so:

$$P(H \text{ or } S) = \frac{1}{4} + \frac{1}{4} = \frac{1}{2}$$

$P(H) \times P(S) = \frac{1}{4} \times \frac{1}{4} = \frac{1}{16}$, but $P(H \text{ and } S) = 0$, so the events

H and S are not independent.

Stop and think   Is it ever possible for two mutually exclusive events to be independent?

Example 17

A card is picked at random from a pack of 52 cards. Use a Venn diagram to find the probability that either an ace or a spade (or both) is picked. Are the events 'ace' and 'spade' independent?

Solution

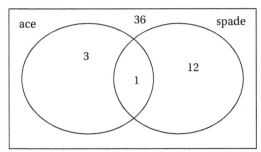

The 52 cards in the pack fall into the areas above. They are

each equally likely to be picked, so $P(A \cup S) = \frac{16}{52} = \frac{4}{13}$.

Note that the events A and S are not mutually exclusive.

$P(A) \times P(S) = \frac{1}{13} \times \frac{1}{4} = \frac{1}{52}$, while $P(A \cap S) = \frac{1}{52}$

So the events S and A are independent.

Stop and think   If events A and B are independent, then A and B' are independent too (as are A' and B, and A' and B'). Can you see why?

Example 18

You are choosing your lunch. Choosing your favourite flavour of crisps is event A. Choosing your favourite flavour of squash is event B. Events A and B are independent, and $P(A) = \frac{1}{4}$, $P(B) = \frac{1}{3}$. Find:

a $P(A \cap B)$ **b** $P(A \cap B')$ **c** $P(A' \cap B')$.

Solution

a $P(A \cap B) = P(A) \times P(B) = \frac{1}{4} \times \frac{1}{3} = \frac{1}{12}$

b $P(A \cap B') = P(A) \times P(B') = \frac{1}{4} \times \frac{2}{3} = \frac{1}{6}$

c $P(A' \cap B') = P(A') \times P(B') = \frac{3}{4} \times \frac{2}{3} = \frac{1}{2}$

An alternative method would have been to use a tree diagram.

Tree diagrams

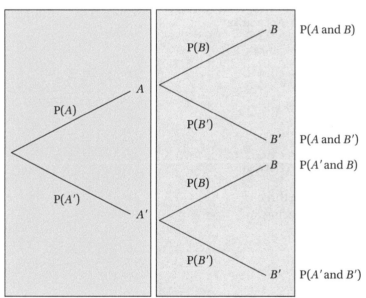

A tree diagram is often the best way to deal with two or more events that may or may not be independent.

The above diagram shows a tree diagram with two events, A and B. The first branch shows the mutually exclusive events A and A', and the second branch shows the mutually exclusive events B and B'. There are four paths through the tree. Note that:

» the probabilities on each section of the branches add up to 1

» to find the probabilities for a sequence of events, you multiply along the branches

» the sum of the probabilities of the end events is always 1.

Example 19

The probability of scoring a strike the first time in ten-pin bowling is 0.21. If a strike is achieved, the probability of another strike is 0.53. However, if a strike was not scored the first time, the probability of getting a strike the second time is 0.39.

a Show this information on a tree diagram.

b Find the probability of getting at least one strike.

Solution

a

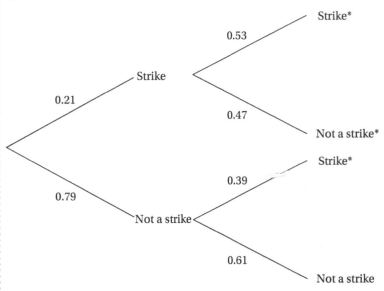

b The event 'at least one strike' is the union of the three events corresponding to the three pathways bearing an asterisk in the tree diagram.

Adding the probabilities for these gives:
$0.21 \times 0.53 + 0.21 \times 0.47 + 0.79 \times 0.39 = 0.5181$

A simpler and faster way of solving this question is to use the complement.

$P(\text{at least 1 strike}) = 1 - P(\text{no strikes}) = 1 - P(0.79 \times 0.61)$
$= 1 - 0.4819 = 0.518$

> The phrase 'at least' in a probability question often suggests that you should use this method.

Example 20

The Venn diagram below represents two events, P and Q. The numbers show how many of the equally likely outcomes correspond to each of the two events.

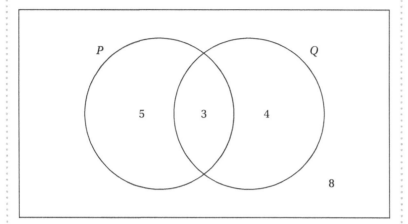

a Use the diagram to find the following probabilities:

   i $P(Q)$ **ii** $P(Q')$ **iii** $P(P \cap Q)$ **iv** $P(P \cup Q)$.

b Are events P and Q independent of each other?

Solution

a **i** Use the Venn diagram to work out the total number of outcomes:

     $5 + 3 + 4 + 8 = 20$ possible outcomes.

     Now use the Venn diagram to work out the number of outcomes in Q.

     $3 + 4 = 7$ outcomes, out of 20 possible outcomes

     So $P(Q) = \dfrac{7}{20}$

  ii Use the Venn diagram to work out the number of outcomes not in Q:

     $5 + 8 = 13$ outcomes out of the 20 possible outcomes.

     So $P(Q') = \dfrac{13}{20}$

 iii This is the probability of P and Q both happening. This is the intersection of the two circles:

     $P(P \cap Q) = \dfrac{3}{20}$

  iv This is the probability of both event P and Q happening, or the union of both events. Add together the numbers in all three sections: $5 + 3 + 4 = 12$.

     $P(P \cup Q) = \dfrac{12}{20}$

b $P(Q) = \frac{7}{20}$ and $P(P) = \frac{8}{20}$

The events P and Q would be independent if $P(P$ and $Q)$ = $P(P) \times P(Q)$.

$P(P \cap Q) = \frac{3}{20}$, but $P(P) \times P(Q) = \frac{56}{400}$

We can see that $\frac{3}{20} \neq \frac{56}{400}$, so P and Q are not independent.

Exercise 2.4A

1 Two events A and event B are mutually exclusive and $P(A) = \frac{1}{5}$ and $P(B) = \frac{1}{4}$. Find:

a $P(A \cap B)$ **b** $P(A \cap B')$ **c** $P(A' \cap B')$

PS 2 Two fair dice are rolled and the results are recorded. Show that the events 'the sum of the scores on the dice is 4' and 'both dice land on the same number' are not mutually exclusive.

3 A bag contains four yellow beads and five blue beads. A bead is chosen at random, the colour recorded but not replaced. A second bead is taken and the colour noted.

Find the probability that:

a the second bead is blue given that the first bead is yellow

b the second bead is yellow given that the first bead is blue

c both beads are yellow

d one yellow bead and one blue bead is chosen.

MM 4 A spinner is split into seven equal sections, with four red sections, two blue and one green. When it lands on a red section, a biased coin with probability $\frac{2}{3}$ of landing on heads is flipped. When the spinner lands on a blue or green section, a fair coin is flipped.

a Draw and label a tree diagram to show the possible outcomes and associated probabilities.

b Danny spins a spinner and flips the appropriate coin. Find the probability that he obtained heads.

C 5 A company sells boxes of eggs. They are of two types, barn (60%) and free-range (40%), and they each come in large and small sizes. 55% of the barn eggs are small, while 75% of the free-range eggs are small. The eggs are put into white and brown boxes to match the colour of the egg.

P(small barn and white) = 0.1, P(small free-range and white) = 0.2, P(large barn and white) = 0.4 and P(large free-range and white) = 0.5.

Find the probability that an egg chosen at random is:

a large and white

b brown

c free-range and brown.

6 Three students, Hamed, Melbin and Zineb, travel independently to college from their homes using one of three methods: walking, cycling or catching the bus.

	Walk	Cycle	Bus
Hamed	0.65	0.1	0.25
Melbin	0.4	0.45	0.15
Zineb	0.25	0.55	0.2

a Calculate the probability that, on any given day:

i all three walk

ii only Zineb catches the bus

iii at least two cycle.

b Tahlia, a friend of Zineb, never travels to college by bus. The probability that Tahlia walks is 0.9 when Zineb walks to college. The probability that Tahlia cycles is 0.7 when Zineb cycles to college.

Calculate the probability that, on any given college day, Zineb and Tahlia travel to college by:

i the same method

ii different methods.

 7 A teacher has a bag of coloured pens for her whiteboard. It contains 3 blue, 7 green and 4 red pens. She picks a pen at random from the bag, does not replace it, and then picks another at random. Find the probability (leaving your answers as fractions in their lowest terms) that she picks:

a a red and a green in some order

b at least one blue pen

c two pens of a different colour.

 8 A company that employs 144 people surveys their staff to see how they get to work. The results say that 51 take a bus and 24 take a train, while 6 employees take both a bus and a train.

a Show that the events 'randomly chosen employee takes the bus' and 'randomly chosen employee takes the train' are not independent.

Now x employees who travel on the train but not the bus start to take the bus as well as the train. It is found that the events 'randomly chosen employee takes the bus' and 'randomly chosen employee takes the train' are now independent.

b Find x.

9 A six-sided die with faces numbered 1 to 6 is rolled. A is the event 'an even number is rolled', B is the event 'an odd number is rolled', while C is the event 'a prime number is rolled'. (Remember that 1 is not a prime number.)

a Which pair of events are mutually exclusive?

The die is biased in such a way that the number 1 is k times as likely as each of the other scores.

b If events B and C are independent, find k.

10 Two events, A and B, are mutually exclusive. Prove that if A and B are also independent, then either $P(A) = 0$ or $P(B) = 0$ or both.

2.5 Conditional probability

You have already met dependent and independent events and seen how the probability of one event happening can depend on whether another event happens or not. With these concepts in mind, you can now look at **conditional probability**. Conditional probability looks at the probability of one event happening given that another event has already occurred.

> The word 'given' in a probability question usually indicates that conditional probability is involved.

In formal notation for conditional probability, a vertical bar stands for 'given'. The probability of event A happening given that event B happens is written $P(A \mid B)$.

For example, you could write the phrase 'the probability of picking a white sock and another white sock' more carefully as 'the probability the second sock picked is white given that the first sock picked is white'. This probability can be written as:

$$P(\text{second sock is white} \mid \text{first sock is white}).$$

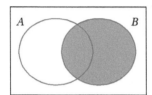

You know that event B has already happened.

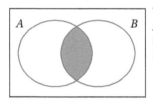

You want the probability of event A, knowing that B has happened. So we are interested in the $A \cap B$ area of the Venn diagram.

If you are looking for the probability that A happens given that B has happened, then the areas in the Venn diagrams above are useful to you. You can see that:

$$P(A \mid B) = \frac{P(A \cap B)}{P(B)}$$

> We can rearrange this formula to read $P(A \mid B) P(B) = P(A \cap B)$.

Stop and think What would this formula look like for the probability of B given that A has happened?

If A and B are independent, the probability of A is unaffected by whether B happens or not. In this case, $P(A \mid B) = P(A)$. In other words, the probability that A happens given that B happens is simply the probability that A happens. Using your formula for conditional probability, you have in this case $P(A) = \frac{P(A \cap B)}{P(B)}$. Rearranging this gives:

$$P(A \cap B) = P(A) \times P(B)$$

KEY INFORMATION

$P(A \mid B)$ means the probability that A occurs given that B has occurred. The formula for conditional probability says:

$$P(A \mid B) = \frac{P(A \cap B)}{P(B)}.$$

A and B are independent if and only if $P(A \mid B) = P(A)$. This implies that events A and B are independent if and only if $P(A \cap B) = P(A) \times P(B)$.

Example 21

J and M are two events such that $P(J) = 0.6$, $P(M) = 0.2$ and $P(M \mid J) = 0.3$.

Find:

a $P(J \mid M)$ **b** $P(J' \cap M')$ **c** $P(M' \cap J)$

d Are J and M independent?

Solution

First find the missing values for the Venn diagram.

$$P(J \cap M) = P(M \mid J) \times P(J) = 0.3 \times 0.6$$
$$= 0.18$$

Now draw a Venn diagram:

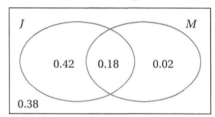

a $P(J \mid M) = \dfrac{P(J \cap M)}{P(M)} = \dfrac{0.18}{0.2}$
  $= 0.9$

b $P(J' \cap M') = 0.38$ (the intersection of not J and not M).

c $P(M' \cap J) = 0.42$ (the intersection of not M with J).

d J and M are independent if and only if $P(M \mid J) = P(M)$, which is untrue, so the events J and M are not independent.

> An alternative method here is to say that:
> $P(J) \times P(M) = 0.6 \times 0.2 = 0.12$
> but $P(J \cap M) = 0.18$. Since $P(J) \times P(M) \neq P(J \cap M)$,
> J and M are not independent.

Example 22

On a day in July you can get sunburnt if the ultraviolet index is high. Event J is the probability of being sunburnt and event K is the probability of a high ultraviolet index, $P(J) = 0.76$, $P(K) = 0.35$ and $P(K \cap J) = 0.25$.

a Find $P(J \mid K)$.   **b** Find $P(K \mid J)$.

Solution

a Using $P(J \mid K) = \dfrac{P(J \cap K)}{P(K)}$ gives $P(J \mid K) = \dfrac{0.25}{0.35} = 0.714$.

  This means that there is a 0.714 chance of getting sunburnt given that it is a day with a high ultraviolet index.

b Using $P(K \mid J) = \dfrac{P(K \cap J)}{P(J)}$ gives $P(K \mid J) = \dfrac{0.25}{0.76} = 0.329$.

  This means that there is a 0.329 chance of it being a high ultraviolet index day given that you have been sunburnt.

Example 23

To qualify for a competition you have to complete two tasks. The probability of completing the first and the second task is 0.53, and the probability of completing the first task is 0.71. What is the probability of completing the second task given that you complete the first?

Solution

P(complete task 2 | complete task 1)

$= \dfrac{\text{P(complete task 1 and complete task 2)}}{\text{P(complete task 1)}}$

$= \dfrac{0.53}{0.71} = 0.746$

Example 24

Nola either cycles or walks to college. The probability that she walks is 0.61. If she cycles, the probability that she will be late is 0.3. If she walks, the probability that she will be late is 0.21.

a Draw a tree diagram to represent the events and their corresponding probabilities.

b Find the probability that Nola has walked given that she was late.

Solution

a

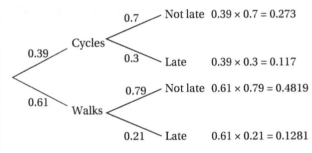

$$
\begin{array}{lll}
0.7 & \text{Not late} & 0.39 \times 0.7 = 0.273 \\
0.3 & \text{Late} & 0.39 \times 0.3 = 0.117 \\
0.79 & \text{Not late} & 0.61 \times 0.79 = 0.4819 \\
0.21 & \text{Late} & 0.61 \times 0.21 = 0.1281
\end{array}
$$

Cycles 0.39, Walks 0.61

b The probability of being late is $(0.39 \times 0.3) + (0.61 \times 0.21)$
$= 0.2451$.
The probability of walking and being late is 0.61×0.21
$= 0.1281$.

Use the conditional probability formula to find the probability of walking given that Nola is late:

$$P(W \mid L) = \frac{P(W \cap L)}{P(L)} = \frac{0.1281}{0.2451} = 0.523$$

Example 25

For events F and G, $P(F) = 0.3$, $P(G \mid F) = 0.42$ and $P(G' \mid F') = 0.61$.

a Draw a tree diagram representing the events and the corresponding probabilities.

b Find $P(G)$.

c Find $P(F' \mid G')$.

Solution

a You will need two sets of branches, one for event F and another for event G. As you already know $P(F)$, put event F on the first set of branches.

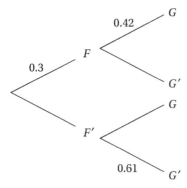

Use the fact that each set of branches must add up to 1 to write in the missing probabilities.

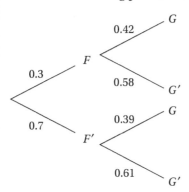

b $P(G) = 0.3 \times 0.42 + 0.7 \times 0.39 = 0.399$

c $P(F' \mid G') = \dfrac{P(F' \cap G')}{P(G')}$

$P(G') = 1 - 0.399 = 0.601$

$P(F' \cap G') = 0.7 \times 0.61 = 0.427$

$P(F' \mid G') = \dfrac{0.427}{0.601} = 0.711$

Exercise 2.5A

1 A card is drawn at random from a pack of 52 playing cards. Given that the card is a spade, find the probability that the card is a jack.

2 In May 2015, the probability of moderate winds in Johor Bahru was $\frac{14}{31}$ and the probability of both moderate winds and humidity above 95% was $\frac{6}{31}$. Find the probability of more than 95% humidity given moderate winds.

3 Two fair coins are flipped and their results recorded. Given that at least one coin lands on tails, find the probability of:

   a two tails   **b** a head and a tail.

4 M and P are two events such that $P(M) = 0.6$, $P(P) = 0.5$ and $P(M \cap P) = 0.4$. Find:

   a $P(M \cup P)$ **b** $P(P \mid M)$

   c $P(M \mid P)$ **d** $P(M \mid P')$.

5 Let E and F be events such that $P(E) = \frac{1}{4}$, $P(F) = \frac{1}{2}$ and $P(E \cup F) = \frac{3}{5}$. Find:

   a $P(E \mid F)$ **b** $P(E' \cap F)$ **c** $P(E' \cap F')$.

6 A local supermarket sells peppers. Bag X contains 12 red and yellow peppers in equal quantities. Another bag, Y, contains four peppers of which three are red and one is yellow.

A pepper is taken out at random from bag X and placed in bag Y. This process is repeated. Finally, a third pepper is taken out of bag Y from the six peppers now inside and the colour noted.

     Event A – when the two peppers taken from bag X are the same colour.

     Event B – when the pepper drawn from bag Y is red.

   a Complete a tree diagram to show all the probabilities.

   b Find $P(B)$.

   c Calculate $P(A \cap B)$.

   d Find $P(A \cup B)$.

   e Find the probability that all the peppers are red given that they are the same colour.

7 A game consists of spinning a spinner and then flipping a coin. The spinner has two colours: red and blue. The probability of spinning red is $\frac{2}{9}$ and the probability of spinning blue is $\frac{7}{9}$.

When the spinner lands on red, a biased coin with a probability of $\frac{3}{5}$ of landing on heads is flipped.

When the spinner lands on blue, a biased coin with a probability of $\frac{1}{5}$ of landing on heads is flipped.

   a Complete a tree diagram to show all the probabilities.

   Mirunalini plays the game.

   b Find the probability that she obtains tails.

c Given that Hilary played the game and obtained a head when she flipped the coin, find the probability that Hilary's spinner landed on blue.

Mirunalini and Hilary play the game again.

d Find the probability that the spinners land on the same colour for both girls.

8 A footballer is practising free kicks. If it is windy, she scores a goal with probability 0.3. If it is not windy, her probability is 0.5. The probability that it is windy is 0.1.

a Draw a tree diagram to show these possibilities.

Find the probability that:

b she scores a goal

c it is windy given that she scores a goal.

9 For an event A:

a what is $P(A|A)$

b what is $P(A|A')$?

Two events A and B are such that $P(A|B) = 0.3$, $P(B|A) = 0.4$.

If $P(A) = 0.5$, find:

c $P(A \cap B)$

d $P(B)$

e $P(A|B')$.

10 For two events R and S, $P(R) + P(S) = 0.8$, while $\dfrac{1}{P(R \mid S)} + \dfrac{1}{P(S \mid R)} = 4$.

a Find $P(R \cap S)$

If $5P(R|S) = 3P(S|R)$:

b find $P(R)$ and $P(S)$.

Mathematics in life and work: Group discussion

You are an actuary working in the area of healthcare and you are asked about a particular test for a disease. The chance of someone having the disease is low: it is $\dfrac{1}{200}$. If someone has the disease, the chance that the test diagnoses this correctly is $\dfrac{99}{100}$. If someone does not have the disease, the chance that the test makes a mistake and says that they do have the disease is $\dfrac{1}{100}$.

1 What is the chance that someone actually has the disease given that they test positive? Use a tree diagram to answer this question.

2 How could you change the probabilities on your tree diagram so that there is a probability of 0.5 that someone has the disease given that they test positive?

3 You can now see why the results of a test like the one above could be termed 'false positive' and 'false negative'. In your group, try varying the probabilities in your tree diagram, and discuss the possible outcomes.

SUMMARY OF KEY POINTS

> The number $n \times (n - 1) \times \ldots \times 2 \times 1$ (for a positive whole number n) is called n factorial, or $n!$. In general, the number of ways of arranging a list of n objects without repetition is $n!$.

> A permutation has the notation ${}^{n}P_{r}$ and describes the number of ways of choosing r objects from n objects if order matters.

> A combination has the notation ${}^{n}C_{r}$ or $\binom{n}{r}$ and describes the number of ways of choosing r objects from n objects if order does not matter.

> ${}^{n}P_{r} = \dfrac{n!}{(n - r)!} \quad {}^{n}C_{r} = \dfrac{n!}{r!(n - r)!}$

> Drawing out the first few rows of Pascal's Triangle can be a useful way to find ${}^{n}C_{r}$ if n is small.

> The probability of an event A, P(A), is a number such that $0 \leqslant P(A) \leqslant 1$.

> If P(A) = 0, A never happens, whereas if P(A) = 1, A always happens.

> The complement of an event A is A', the event A does not happen, and P(A') = 1 − P(A).

> P($A \cup B$) means the probability that A or B or both occur (A union B).

> P($A \cap B$) means the probability that both A and B occur (A intersection B).

> If A and B are mutually exclusive they cannot occur together, which means P($A \cap B$) = 0 and P($A \cup B$) = P(A) + P(B).

> P($A \mid B$) means the probability that A occurs given that B has occurred.

> The formula for conditional probability is $P(A \mid B) = \dfrac{P(A \cap B)}{P(B)}$.

> A and B are independent if and only if P($A \mid B$) = P(A), which implies that the events A and B are independent if and only if P($A \cap B$) = P(A) × P(B).

Exam-style questions

1 There are two ways to watch a movie in a hotel: digital download or DVD. The number of films in each of four subject categories is shown in the table.

	Crime	Romantic comedy	Science fiction	Documentary	Total
DVD	8	16	18	18	**60**
Digitial download	16	40	14	30	**100**
Total	**24**	**56**	**32**	**48**	**160**

A film is selected at random. Calculate the probability that the film is:

a not a romantic comedy

b either crime or science fiction

c a digital download given that it is a documentary.

2 There are three palm trees in a garden, and they are harvested for three, four and five coconuts, respectively. The coconuts from each tree can be identified.

a Bella makes a selection of seven coconuts, including at least two from each tree. How many different selections can she make?

b How many arrangements of all the coconuts in a row are there if the three coconuts from the first tree have to be together?

3 Uche, Eli and Ellis play darts. The independent probabilities that Uche, Eli and Ellis hit a bull's eye are 0.62, 0.17 and 0.68, respectively.

Find the probability that:

a none of them hits a bull's eye

b at least one hits a bull's eye

c two of them hit a bull's eye given that at least one of them hits the bull's eye.

4 There are 185 students at a youth club. Students can choose three activities: 112 students take climbing, 70 take table tennis, 81 take bowling, 35 take climbing and bowling, 16 take table tennis and bowling, 40 take climbing and table tennis, and 11 take all three activities.

a Draw a Venn diagram to represent this information.

A student is chosen at random. Find the probability that this student:

b does none of the activities

c does bowling only.

d What is the probability that the student does none of the activities given that she or he does not take bowling?

5 Kath is sitting a multiple choice paper. The probability that she answers a question slowly (> 1 min) and gets it right is $\frac{13}{24}$. The probability that she answers quickly ($\leq$1 min) and gets it wrong is $\frac{1}{8}$.

	Time taken $\leq$ 1 min	Time taken > 1 min	Total
Question right		$\frac{13}{24}$	
Question wrong			
Total	$\frac{1}{4}$		1

a Complete the remaining cells in the table.

b Given that Kath gets a question right, find the probability that she answered it slowly.

PS **6** A karting venue has two red, one blue and one green kart. Two karts are to be raced at random.

   a Draw a tree diagram to illustrate all the possible outcomes and associated probabilities.

   b Find the probability that a blue kart and a green kart are raced.

   c A second race takes place. This time, the chance of the first kart chosen being green is x, and the chance of the first kart chosen being blue is $2x$. The second kart is chosen at random from the remaining karts. If the probability of a green kart and a red kart racing is $\frac{1}{5}$, what is x?

C **7** In a group of 32 students, 24 take art and 16 take music. One student takes neither art nor music. The Venn diagram below shows the events art and music. The values p, q, r and s represent numbers of students.

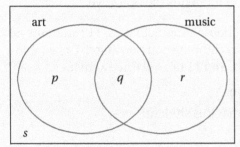

   a Write down the values of p, q, r and s.

   b A student is selected at random. Given that the student takes music, write down the probability that the student takes art.

   c Hence, show that taking music and taking art are not independent events.

   d Two students are selected at random, one after the other. Find the probability that the first student takes only music and the second student takes only art.

PS **8** Farhan is a juggler. In his box are six different green balls, three different red balls, and two different blue balls.

   a He picks four balls, and his choice includes at least one of each colour. How many different choices could he make?

To start his act, Farhan arranges all eleven balls in a straight line.

   b How many arrangements of the balls (as seen by the audience) are there if there is a blue ball on each end?

   c Farhan now removes the blue balls and arranges the rest in a line. How many arrangements of the balls (as seen by the audience) are there if there is a green ball on each end?

C **9** In a class of 100 boys, 47 boys cycle and 62 boys swim. Each boy must do at least one sport from cycling and swimming.

   a Write down the number of boys who both cycle and swim.

One boy is selected at random.

b Find the probability that he does only one sport.

c Given that the boy selected does only one sport, find the probability that he cycles.

Let C be the event that a boy cycles and S be the event that a boy swims.

d Explain why C and S are not mutually exclusive.

e Show that C and S are not independent.

PS **10** All the kings, queens and jacks are taken from a pack of playing cards, and they are placed in a line at random.

a What is the probability that the four jacks are together?

b What is the probability that two kings are at one end and the other two kings are at the other end?

c You choose a hand of four cards from the 12 you have taken from the pack, so that you have at least one king, at least one queen and at least one jack. How many possible hands are available to you?

d If you choose your four cards at random, what is the probability that you have at least one king, at least one queen and at least one jack in your hand?

C **11** Two players each choose one of three spinners in turn. Each player spins his or her chosen spinner and the one showing the higher number is the winner.

The faces of the three spinners are labelled as follows:

Spinner 1	10, 10, 10, 20, 20, 30
Spinner 2	5, 11, 15, 21, 48
Spinner 3	1, 9, 14, 19, 57

The total on each spinner is 100, so the game would appear fair. Do you agree? Explain your answer.

MM **12** A new business is setting up to make premium chocolates.

Machine J makes 25% of the chocolates.

Machine K makes 45% of the chocolates.

Machine L makes the remaining chocolates.

It is known that 2% of the chocolates made by machine J are misshapen, 3% of the chocolates made by machine K are misshapen and 5% of the chocolates made by machine L are misshapen.

a Draw a tree diagram to illustrate all the possible outcomes and associated probabilities.

A chocolate is selected at random.

b Calculate the probability that the chocolate is made by machine J and is not misshapen.

c Calculate the probability that the chocolate is misshapen.

d Given that the chocolate is misshapen, find the probability that it was not made by machine K.

C **13** All the number cards from 2 to 10 in a pack of standard cards are removed to form a new pack of 36 cards. The new pack is shuffled, and you pick a card.

    a What is the probability that you pick a card bearing an even number?

You pick a second card. Event S is 'the two cards add up to an even number'. Event T is 'the two cards multiply to an odd number'.

    b Are S and T mutually exclusive? Justify your answer.

    c Are S and T independent? Again, justify your answer.

PS **14** James writes each of the digits 1, 2, 3, 4, 5, 6, 7, 8 and 9 on identical pieces of card.

    a How many different five-digit numbers can he make from these digits if repeats are not allowed?

    b A number is divisible by 9 if and only if its digits add to a multiple of 9. If James picks three digits at random from his 9 to create a three-digit number, what is the probability the number will be a multiple of 9?

James then arranges his digits into three groups, numbers divisible by 3, numbers that have a remainder 1 when divided by 3, and numbers that have a remainder 2 when divided by 3.

    c If James picks four numbers (including at least one from each group), in how many ways can he do this?

    d James now picks one card from each group at random, and then one more card at random. What is the probability that the sum of his cards is divisible by 3?

MM **15** Ding and his neighbour Li work at the same place.

On any day Ding travels to work, he uses one of three options: his car only, a tram only or both his car and a tram. The probability that he uses his car, either on its own or with a tram, is 0.8. The probability that he uses both his car and a tram is 0.36.

Calculate the probability that, on any particular day when Ding travels to work, he:

    a uses his car only

    b uses a tram.

On any day, the probability that Li travels to work with Ding is 0.56 when Ding uses his car only, 0.66 when Ding uses both his car and a tram and 0.25 when he uses a tram only.

    c Calculate the probability that, on any particular day when Ding travels to work, Li travels with him.

    d Assuming that the choices are independent from day to day, calculate, to three decimal places, the probability that during any particular week of five days when Ding travels to work every day, Li never travels with him.

C 16 In a school where 48% of the students are boys and 52% are girls, 57% of the students play cricket. $54\frac{1}{6}$% of the boys play cricket. A student is chosen at random. Find the probability that the student:

a is a boy who plays cricket

b is a girl who does not play cricket

c is a boy given that the student plays cricket

d are the events 'chosen student plays cricket' and 'chosen student is a girl' independent?

MM 17 Evan has eggs every day, either scrambled (35% of the time) or poached (40% of the time) or fried (25% of the time). If his eggs are scrambled, he has them on toast 70% of the time, while poached eggs are on toast 75% of the time and fried eggs are on toast 25% of the time.

a Draw a tree diagram to show the possibilities.

b What is the probability that he has his eggs on toast?

c What is the probability that he chose fried eggs given that they were on toast?

PS 18 A red die is rolled with result R, and a green die is rolled with result G. The results are multiplied together to obtain RG.

a Draw a sample space for the possible outcomes.

b Use this to find P(RG is a square number).

c Find P(RG is even given that it is a square).

d Calculate P(RG is odd given that it is a square).

C 19 In the staff–student liaison group at a school, there are 24 students and five members of staff. A new committee of five members is formed. Find out how many ways this may be done if:

a it contains three members of staff and two students

b it contains more students than staff but at least one staff member.

MM 20 A helicopter has 12 passenger seats arranged into four rows and three columns, as shown.

Ten passengers including Alan and Tina take their seats.

a If the two empty seats can be anywhere, how many passenger arrangements are there?

b If Alan has to sit at the end of a row and Tina must not sit at the end of a row, how many passenger arrangements are there?

   c If the central two seats must be the empty ones, and Alan and Tina must sit side by side in a row, how many passenger arrangements are there?

   d A passenger arrangement is picked at random. Expressed as a fraction, what is the probability that the central two seats are the empty ones, and that Alan and Tina sit side by side in a row?

(Answers can be left in factorial form for all four parts.)

(c) 21 Chyou is rolling two biased dice: one red, one green. The chance of her getting an even number on the red is $\frac{7}{12}$, while the chance of her getting an even number on the green is $\frac{1}{2}$. The chance that she rolls an odd number on both is $\frac{1}{4}$. Some of this information is shown in the table below.

	Red even	Red odd	
Green even			$\frac{1}{2}$
Green odd		$\frac{1}{4}$	
			1

   a Copy and complete the table.

   b Given that the sum of her two numbers is even, find the probability that the product of the two numbers is also even.

Mathematics in life and work

You have been asked to suggest to an insurance company what their premium should be for a policy that protects a car owner against two big repair bills, for the exhaust and the gearbox.

The probability that the exhaust fails in the next year is 0.1, while the probability that the gearbox fails is 0.05.

If the gearbox fails, the cost is $3000, while if the exhaust fails, the cost is $500.

1 Draw a tree diagram to show these probabilities and the costs of each of the four possible outcomes for the motorist.

2 The expected cost to the motorist is given by multiplying the probability of each outcome by its cost, and then adding the four results. What is the expected cost to the motorist?

3 The insurance company needs to expect to make a profit of $20 on each car. What premium should the motorist be charged?

3 DISCRETE RANDOM VARIABLES

Mathematics in life and work

As you saw in **Chapter 2 Probability, permutations and combinations**, there are many situations where using probability can help you to make decisions or know what to expect. A random variable can be used to model outcomes of trials or experiments. These can be formal experiments such as those performed in chemistry lessons or informal experiments such as trying to score a penalty in football. Probability allows us to work out the chance of these outcomes happening or what we would expect to happen.

The statistical distributions in this chapter are relevant in a range of careers. For example:

> If you were a quality control officer in a factory, you could use the binomial distribution to identify whether the number of defective products is too high.

> If you were mechanical engineer, you could use the geometric distribution to predict how many items a machine can make before it makes a faulty item. This can be used when making a decision about when the machine should be maintained or replaced.

> If you were traffic safety expert, expectation and variance could be used to simulate traffic patterns and inform decisions to improve safety.

LEARNING OBJECTIVES

You will learn how to:

> construct a probability distribution table for a discrete random variable X

> calculate the expectation, $E(X)$, and variance, $Var(X)$, of a discrete random variable

> calculate binomial probabilities using the notation $X \sim B(n, p)$

> calculate expectation and variance for a binomial distribution

> calculate geometric probabilities using the notation $X \sim Geo(x)$

> calculate expectation of a geometric distribution

> recognise practical situations where these distributions are suitable models.

LANGUAGE OF MATHEMATICS

Key words and phrases you will meet in this chapter:

> binomial distribution, binomial coefficient, continuous data, discrete data, discrete random variable, expectation, failure, geometric distribution, independent, probability distribution, random variable, success, trial, uniform distribution, variance

PREREQUISITE KNOWLEDGE

You should already know how to:

› complete tables and grids to show outcomes and probabilities

› apply probability laws

› calculate the mean and variance from a frequency table

› understand the notation ^nC_r and $n!$

You should be able to complete the following questions correctly:

1 A fair coin is flipped three times and heads or tails is recorded each time.

   a Write down a sample space for the recorded outcomes.

   b Find the probability that exactly one head is recorded.

   c Find the probability that at least two tails are recorded.

2 Find the probability that:

   a you get a 3 or a 5 when you roll a fair die

   b you get a 1 and a 5 when 2 fair dice are rolled.

3 Calculate the mean and the variance of the variable X in the following frequency table.

Number of days absent (x)	0	1	2	3	4
Frequency	12	8	7	1	2

4 Calculate:

   a 6!

   b $^{10}C_4$

3.1 Probability distributions of discrete random variables

From your secondary maths course and **Chapter 1 Representation of data**, you will remember that **discrete data** can only take certain values, whereas **continuous data** can take values within a range.

A **discrete random variable** is a **random variable** that can only take particular values. It is usually represented by an uppercase letter such as X. The particular values that X can take are represented by the lowercase letter x. A **probability distribution** shows all the possible values a discrete random variable can take, plus the probability of each value occurring.

A probability distribution can be defined in three forms:

1 a table

2 a graph

3 a simple function.

> You need to be able to construct a probability distribution table.

The probabilities of all the possible values that a discrete random variable can take add up to 1.

For a discrete random variable X:

$$\sum_x P(X = x) = 1$$

Flipping a fair coin twice and counting the number of heads would mean event X ('the number of heads') and the values it can take, x, could be 0, 1 or 2. You can check that this works using the formula:

$$\sum_{x=0,1,2} P(X = x) = 1 = P(X = 0) + P(X = 1) + P(X = 2)$$

$$= \frac{1}{4} + \frac{1}{2} + \frac{1}{4} = 1$$

$P(X = x)$ is the notation for probability that the random variable, X, has the value x. It can also be written as $P(x)$.

KEY INFORMATION

The probabilities of all the possible values that a discrete random variable can take add up to 1.

Example 1

Construct the probability distribution table for X, where X is the score on a fair, six-sided die:

a in the form of a table

b in the form of a graph.

Solution

X is the random variable representing the score when a die is rolled, and x represents the values X can take.

a List all the values X can take in the table:

x	1	2	3	4	5	6
$P(X = x)$	$\frac{1}{6}$	$\frac{1}{6}$	$\frac{1}{6}$	$\frac{1}{6}$	$\frac{1}{6}$	$\frac{1}{6}$

The corresponding probabilities are entered underneath each variable. It is always useful to check that they add up to 1.

b The probability of each outcome is plotted on a graph as shown below.

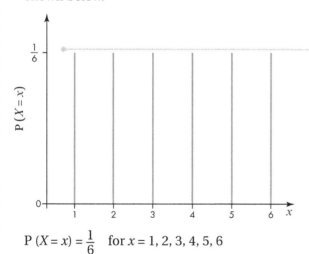

$P(X = x) = \frac{1}{6}$ for $x = 1, 2, 3, 4, 5, 6$

The graph shows that every outcome is equally likely to happen. This is called a discrete **uniform distribution**. The probability distribution of a random variable with a uniform distribution always has the same shape.

Stop and think   How can you tell that this is a discrete uniform distribution?

The probability distribution is defined as a constant with a discrete outcome.

Example 2

There are two fair spinners. One spinner is split into four equal sections with the numbers 0–3. The other spinner has three equal sections with the numbers 1–3. The random variable X is sum of the scores on the two spinners. Find the probability distribution of X.

Solution

First you need to create a sample space of all the possible outcomes of the spinners. These are written as (spinner 1, spinner 2):

```
0,1  1,1  2,1  3,1
0,2  1,2  2,2  3,2
0,3  1,3  2,3  3,3
```

You can now work out the corresponding values for X.

```
1   2   3   4
2   3   4   5
3   4   5   6
```

This allows you to find the probability of each value of the random variable.

x	1	2	3	4	5	6
$P(X = x)$	$\frac{1}{12}$	$\frac{2}{12}$	$\frac{3}{12}$	$\frac{3}{12}$	$\frac{2}{12}$	$\frac{1}{12}$

Check that your probabilities add up to 1.

In **Example 2**, every value of X is not equally likely. The graph of the probability distribution would look like this:

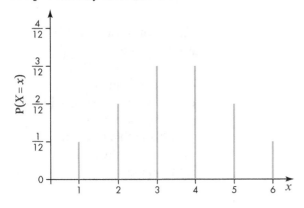

Example 3

The discrete random variable X has the following probability distribution:

x	0	1	2	3	4
$P(X = x)$	0.1	a	0.3	a	0.1

Find:

a the value of a

b $P(1 \leqslant X < 3)$

c the mode.

Solution

a Use $\sum\limits_{x} P(X = x) = 1$

$0.1 + a + 0.3 + a + 0.1 = 1$

$2a + 0.5 = 1$

$2a = 0.5$

$a = 0.25$

b This is asking for the probability between 1 and 3, including 1 but not including 3. In other words, find the probability that $X = 1$ or $X = 2$.

$P(1 \leqslant X < 3) = P(X = 1) + P(X = 2)$

$= 0.25 + 0.3$

$= 0.55$

> As the events are mutually exclusive, you add the probabilities.

> You can also use the fact that:
> $P(1 \leqslant X < 3) = 1 - P(X = 0) - P(X = 3) - P(X = 4) = 1 - 0.1 - 0.25 - 0.1 = 0.55$

c The mode is the most likely (most frequent) value. The highest probability is 0.3, so the mode is 2.

Example 4

A discrete random variable, X, can take the values 1, 2, 3 and 5 with probability distribution:

$$P(X = x) = \begin{cases} 2k & \text{for } x = 1 \\ k & \text{for } x = 2, 3 \text{ and } 5 \\ 0 & \text{otherwise} \end{cases}$$

> The probability distribution of the random variable is given in a form of simple function.

a Write down the probability distribution in the form of a table.

b Find the value of k.

c Find $P(X \leqslant 3)$.

Solution

a The probability distribution table is as follows.

x	1	2	3	5
$P(X = x)$	$2k$	k	k	k

b Since X is a random variable, $\sum_{\text{all } x} P(X = x) = 1$

Therefore, $2k + k + k + k = 1$

$5k = 1$

$k = 0.2$

c $P(X \leqslant 3) = P(X = 1) + P(X = 2) + P(X = 3)$

$\qquad = 2k + k + k$

$\qquad = 4 \times 0.2$

$\qquad = 0.8$

> From part **b**, you already worked out that $k = 0.2$.

Another way to think about this problem is that the probability that 3 or less is happening is the same as the probability that 5 is not happening.

You can write:

$P(X \leqslant 3) = 1 - P(X = 5)$

$\qquad = 1 - 0.2$

$\qquad = 0.8$

> **KEY INFORMATION**
> There are three different forms to define a probability distribution:
>
> **1** a table
> **2** a graph
> **3** a simple function.

Exercise 3.1A

(C) 1 The random variable X is given by the sum of the scores when two fair dice are rolled.

   a Find the probability distribution of X.

   b Draw a diagram to show the distribution and describe the shape.

2 The random variable A is given by the product of the scores when two ordinary dice are rolled. Find the probability distribution of A.

3 The random variable Y has the probability distribution:

y	−1	0	1	2	3
$P(Y = y)$	0.1	0.2	a	0.25	0.3

   a Find $P(Y = 1)$.

   b Find $P(Y \leqslant 0)$.

   c Find $P(Y > 1)$.

(C) **Communication**　　(MM) **Mathematical modelling**　　(PS) **Problem solving**

4 Two tetrahedral dice labelled 1, 2, 3 and 4 are rolled. The random variable X represents the sum of the numbers shown on the dice.

a Find the probability distribution of X.

b Calculate the probability that any throw of the dice results in the value of X being a prime number.

5 The random variable X has the probability distribution:

x	2	3	5	7	11	13
$P(X = x)$	k	k	k	k	k	k

a Find k.

b Find $P(X < 7)$.

c Find $P(X \geqslant 3)$.

d Find $P(2 \leqslant X < 11)$.

6 There are three blue counters and six green counters in a bag. A counter is picked out of the bag, the colour noted and then replaced three times. The random variable X is the number of blue counters taken from the bag.

a Find the probability distribution of X.

b Find the probability of getting one or more blue counters.

7 A biased die is thrown and X is the number on the top face. The probability that the die lands on a particular number is proportional to that value.

Find:

a the probability distribution of X

b $P(2 \leqslant X < 4)$

c $P(\text{even number})$.

8 The discrete random variable R has the probability distribution:

$$P(R = r) = \begin{cases} 0.1r & \text{for } r = 1, 2, 3 \\ k & \text{for } r = 4 \\ 0 & \text{otherwise} \end{cases}$$

Construct a probability distribution table for R and find the value of k.

9 Professor Han has appointments with four undergraduates every Friday morning. Experience suggests that on each Friday the number of these undergraduates who were late for the appointments with him may be modelled by the random variable X. The probability distribution of X is given in the following table:

x	0	1	2	3	4
$P(X = x)$	0.4	0.25	$4k$	$2k$	k

Calculate the probability that on a Friday morning at least two under graduates are late.

10 The probability distribution for the number of cars, C, involved in each minor accident on a particular road can be modelled as follows:

c	0	1	2	3	4	5
$P(C = c)$	0	$3r$	$9r$	$4r$	$3r$	r

a Find the value of r.

b A traffic officer suggests that the probability distribution for the number of cars, C_2, involved in each minor accident is given as follows:

$$P(C_2 = c) = \begin{cases} (0.25 + c)k & \text{for } c = 0, 1, 2, 3 \\ (c - 0.5)k & \text{for } c = 4, 5 \\ 0 & \text{otherwise} \end{cases}$$

Construct a probability distribution table for this new model and find the probability that there is at least one car involved in each minor accident.

3.2 Expectation of X

The **expectation** of a random variable X is the long-run average value. In other words, it demonstrates the expected outcome when a **trial** is repeated a large number of times. The expectation of X is normally written as $E(X)$ or mean.

A fair die is rolled 100 times and the score is recorded in the table below:

X	1	2	3	4	5	6
Frequency	19	19	12	17	18	15

You can use the formula:

$$\text{mean} = \frac{\sum xf}{\sum f}$$

to calculate the mean score of this data, as you did in **Chapter 1 Representation of data**.

$$= \frac{1 \times 19 + 2 \times 19 + 3 \times 12 + 4 \times 17 + 5 \times 18 + 6 \times 15}{19 + 19 + 12 + 17 + 18 + 15}$$

$$= 3.41$$

Stop and think How many rolls of the die might be needed for the relative frequency to yield probabilities that are very close to the theoretical probabilities?

As the number of times the die is rolled increases, the relative frequency tends to the theoretical probability. This would result in the probability distribution below:

X	1	2	3	4	5	6
p	$\frac{1}{6}$	$\frac{1}{6}$	$\frac{1}{6}$	$\frac{1}{6}$	$\frac{1}{6}$	$\frac{1}{6}$

You can find the mean score of the probability distribution table above:

$$E(X) = \frac{\sum xp}{\sum p}$$

$$= \frac{1 \times \frac{1}{6} + 2 \times \frac{1}{6} + 3 \times \frac{1}{6} + 4 \times \frac{1}{6} + 5 \times \frac{1}{6} + 6 \times \frac{1}{6}}{\frac{1}{6} + \frac{1}{6} + \frac{1}{6} + \frac{1}{6} + \frac{1}{6} + \frac{1}{6}}$$

$$= 3.5$$

The sum of probabilities in a probability distribution must add up to 1. This means we can rewrite the formula:

$$E(X) = \sum xP(x)$$

KEY INFORMATION

The expectation of a random variable is given by the formula $E(X) = \sum xp$.

Example 5

A survey was taken of 100 randomly chosen households in which people were asked how many computers they owned. The results are shown in the table.

Number of computers	0	1	2	3
Frequency	17	42	31	10

Let X be the random variable that represents the number of computers in a household.

a Find the probability distribution of X.

b Find the expectation of the random variable X.

Solution

a

x	0	1	2	3
$P(X = x)$	0.17	0.42	0.31	0.1

$$P(x) = \frac{\text{Frequency}}{\text{Total Frequency}}$$

b $E(X) = \sum xp$

$$= 0 \times 0.17 + 1 \times 0.42 + 2 \times 0.31 + 3 \times 0.1$$

$$= 1.34$$

Example 6

Y is a random variable with the distribution below and has an expectation of 3.

y	1	2	3	4	5
$P(Y = y)$	a	0.3	b	0.2	0.15

a Find the values of a and b.

b Find $E(Y^2)$.

Solution

a $1 \times a + 2 \times 0.3 + 3 \times b + 4 \times 0.2 + 5 \times 0.15 = 3$ ......●------- $E(Y) = \Sigma yp$

$$a + 3b = 3 - (0.6 + 0.8 + 0.75)$$

$$a + 3b = 0.85 \qquad\qquad [1]$$

$a + 0.3 + b + 0.2 + 0.15 = 1$ ......●------- All probabilities add up to 1.

$$a + b = 1 - (0.3 + 0.2 + 0.15)$$

$$a + b = 0.35 \qquad\qquad [2]$$

$$2b = 0.5$$ ......●------- [1] – [2] to eliminate a.

$$b = 0.25$$

$$a + 0.25 = 0.35$$ ......●------- Substitute b into [2].

$$a = 0.1$$

b

y	1	2	3	4	5
y^2	1	4	9	16	25
$P(Y = y)$	0.1	0.3	0.25	0.2	0.15

$E(Y^2) = \Sigma y^2 p$

$\quad = 1 \times 0.1 + 4 \times 0.3 + 9 \times 0.25 + 16 \times 0.2 + 25 \times 0.15$

$\quad = 10.5$

> **KEY INFORMATION**
> $E(Y^2) = \Sigma y^2 p$. This is also true for $E(Y^n) = \Sigma y^n p$.

Exercise 3.2A

 Find the expected value for each of the distributions below.

a

x	–2	–1	0	1	2	3
$P(X = x)$	0.13	0.27	0.1	0.18	0.22	0.1

b

y	2	4	6	8
$P(Y = y)$	$\frac{1}{12}$	$\frac{5}{12}$	$\frac{1}{3}$	$\frac{1}{6}$

2

z	5	6	7	8
$P(Z = z)$	0.4	0.3	0.2	0.1

a Find $E(Z)$.

b Find $E(Z^2)$.

c State, with a reason, whether $E(Z^2) = (E(Z))^2$.

3 The random variable D is given by the difference between the largest score and the smallest score when two fair dice are thrown.

a Find the probability distribution of D.

b Find $E(D)$.

4 a

x	0	1	2	3	4
$P(X = x)$	0.2	0.15	0.35	a	0.1

i Find a.

ii Find $E(X)$.

iii Find $E(X^2)$.

b

y	−2	0	2	4
$P(Y = y)$	k	$2k$	$3k$	k

i Find k.

ii Find μ.

iii Find $E(Y^2)$.

> The expectation of Y, $E(Y)$, has the same value as the mean of Y: $\mu = E(x)$.

5 Every day Zarak receives mail. He decides to count how many letters he receives each day on 50 randomly chosen days. His results are shown below.

Number of letters	1	2	3	4	5
Frequency	7	22	18	1	2

By modelling the number of letters as a discrete random variable, find:

a the probability distribution

b $E(X)$

c $P(X > E(X))$.

6 A biased coin, where the probability of landing on heads is 0.3, is flipped three times. If X is the random variable that counts the number of heads, find $E(X)$.

(PS) **7** The random variable X has the distribution shown below and the expectation of X is $\frac{5}{12}$.

x	−1	0	1	2
$P(X = x)$	a	$4b$	$2b$	a

Find a and b.

8 A discrete random variable, Z, can only take the values 4 or 12. If $E(Z) = 7$, find the probability distribution of Z.

9 The random variable, X, has the probability distribution shown below.

$$P(X = x) = \begin{cases} xk & \text{for } x = 1, 4 \\ \dfrac{1}{8} & \text{for } x = 2, 3, 5, 6 \\ 0 & \text{otherwise} \end{cases}$$

a Find the value of k.

b Find $E(X)$.

c Hence find $E(2X)$.

10 The discrete random variable Y is such that $E(Y) = 6$.

a $P(Y = y) = \begin{cases} \dfrac{y}{20} & \text{for } y = 2, 4, 6, 8 \\ 0 & \text{otherwise} \end{cases}$

Construct the probability distribution table for Y.

b Calculate the mean of Y^{-1}.

3.3 Variance of X

The **variance** of a random variable X is a measure of spread of the random variable from the expected value. The variance is normally written as $\text{Var}(X)$ and can also be denoted by σ^2.

The variance of a random variable can be found using the formula :

$$\text{Var}(X) = \sum x^2 p - (E(X))^2$$

This can be thought of as 'the mean of the squares minus the square of the mean'. The standard deviation can be calculated by $\sigma = \sqrt{\text{Var}(X)}$.

> $\sum x^2 p$ can also be written as $E(X^2)$. $(E(X))^2$ can also be written as $E^2(X)$.

Stop and think Why is the variance always a positive value? When the variance is square-rooted, why do we only consider the positive value for the measure of standard deviation?

KEY INFORMATION

The variance of a random variable is given by the formula:
$\text{Var}(X) = \sum x^2 p - (E(X))^2$
or $\text{Var}(X) = E(X^2) - E^2(X)$

Example 7

X is a random variable with the following probability distribution:

x	1	2	3	4
P(X = x)	0.2	0.35	0.3	0.15

a Find E(X).

b Calculate Var(X).

Solution

a $E(X) = \Sigma xp$

$\quad\quad = 1 \times 0.2 + 2 \times 0.35 + 3 \times 0.3 + 4 \times 0.15$

$\quad\quad = 2.4$

b $Var(X) = \Sigma x^2 p - (E(X))^2$

$\quad\quad = 1^2 \times 0.2 + 2^2 \times 0.35 + 3^2 \times 0.3 + 4^2 \times 0.15 - 2.4^2$

$\quad\quad = 0.94$

Exercise 3.3A

1 **a** Find the variance for each of the distributions below.

x	−2	−1	0	1	2	3
P(X = x)	0.13	0.27	0.1	0.18	0.22	0.1

b

y	2	4	6	8
P(Y = y)	$\frac{1}{12}$	$\frac{5}{12}$	$\frac{1}{3}$	$\frac{1}{6}$

2 *D* is the random variable that represents the score on a fair die. Find E(D) and Var(D).

3 Two fair tetrahedral dice are rolled and the random variable *S* is the sum of their scores.

    a Calculate E(S).

    b Calculate Var(S).

4 A random variable, *X*, has the probability distribution shown in the table below. The expectation of *X* is 5.95.

x	2	5	*a*	8
P(X = x)	0.1	*p*	0.2	0.35

    a Find *p*.   **b** Find *a*.   **c** Find Var(X).

5 A random variable Y has a probability distribution shown below and the expectation of Y is 3.8.

y	1	3	5	7
$P(Y = y)$	a	$3b$	$2a$	b

a Find a and b.

b Calculate $\text{Var}(Y)$.

6 A stockbroker has a choice of two stocks in which to invest. He knows that for stock A he will get a return of 10%, 25% or –5% with probabilities 0.6, 0.25, 0.15, respectively. For stock B he will get a return of 10%, 50% or –30% with probabilities 0.55, 0.25, 0.2, respectively.

By modelling the returns as random variables:

a find $E(A)$ and $E(B)$

b find $\text{Var}(A)$ and $\text{Var}(B)$

c comment how these values could affect the decision of the stockbroker.

7 A fair die is rolled until a 1 is seen or until the die has been rolled three times. R is a random variable that represents the number of rolls of the die.

a Show that $P(R = 3) = \dfrac{25}{36}$.

b Find the probability distribution.

c Find $E(R)$ and $\text{Var}(R)$.

8 A school council of 8 boys and 12 girls appoints two of its members to represent it. Assuming that each member is equally likely to be appointed:

a construct a probability distribution table of the number of girls who are appointed

b find the expected number of girls who are appointed.

9 A motor company can manufacture one of two possible new car models. The finance director has been asked to advise which one to choose.

Model A will yield a profit of $3000 with probability 0.5, a profit of $2500 with probability 0.3 and a loss of $500 with probability 0.2.

Model B will yield a profit of $3500 with probability 0.4, a profit of $3000 with probability 0.4 and a loss of $1000 with probability 0.2.

Determine which car model the finance director should support by modelling these two car models as random variables.

10 A fair spinner with only odd numbers 1, 3, 5 is spun.

An unbiased coin is then flipped the number of times indicated by the score on the spinner.

Let H denote the number of heads obtained.

a Show that $P(H = 3) = \dfrac{7}{48}$.

b Construct the probability distribution table of H.

c Find $E(H)$ and $\text{Var}(H)$.

Mathematics in life and work: Group discussion

The expectation and standard deviation often need to be considered together to be able to make decisions about data sets. Suppose that you are the owner of a taxi company and you have received an urgent call from a passenger who would like to be collected from her house and taken to the airport. The passenger needs to be at the airport in 30 minutes or she will miss her flight. You have three taxis in the area that you could assign to this job and, from their previous airport journeys, you know that their means and standard deviations are as follows:

Antonio's taxi:	$\mu = 31$ mins	$\sigma = 5$ mins
Barney's taxi:	$\mu = 29$ mins	$\sigma = 2$ mins
Chloe's taxi:	$\mu = 27$ mins	$\sigma = 7$ mins

How might this data be used to allocate a taxi for this job? Rank the taxis in order of preference and give reasons for your choices.

3.4 Geometric and binomial distributions

The **geometric distribution** and **binomial distribution** are both discrete probability distributions that describe discrete random variables. The geometric distribution is used when we are considering the number of trials until an event of interest happens, while the binomial distribution is used when we have a fixed number of trials.

The geometric distribution

The geometric distribution is about 'success' and 'failure', as there are only two possible outcomes to each trial. For instance, if you flip a coin once it can only land on heads or tails. The 'success' would be heads, while the 'failure' would be tails (or vice versa, as long as you refer to each outcome consistently).

If you were to roll a die repeatedly until you rolled a 6 and you wanted to know the probability that it would take four rolls, then the failure would be rolling any number apart from 6 and the success would be rolling a 6. This would mean that we would have three failures followed by a success, which would have the probability $\left(\frac{5}{6}\right)^3 \times \frac{1}{6}$.

It is important that you can identify situations, which can be modelled using the geometric distribution. You can use the geometric distribution:

- if there are just two possible outcomes to each trial, success and failure, with fixed probabilities of p and q, respectively, where $q = 1 - p$

- if all trials are independent of each other

- if the probability of success, p, is the same in each trial

- if you continue the trials until the first success.

KEY INFORMATION

You can write $X \sim \text{Geo}(p)$.

The discrete random variable X is the number of trials up to and including the first success. X is modelled by the geometric distribution Geo(p).

The probability of success in a single trial is usually taken as p and the probability of failure as q. The probability of achieving exactly one success in the nth trial is shown below.

P(first success on the nth trial) = $p \times q^{n-1}$ or $P_n = p(1-p)^{n-1}$

n = number of trials up to and including the first success

$n - 1$ = number of failures

p = probability of success in one trial

$q = 1 - p$ = probability of failure in one trial

KEY INFORMATION

If $X \sim$ Geo(p), then
$P_n = p(1-p)^{n-1}$.

Example 8

Thanh is practising taking penalties. The probability that he scores a penalty is 0.3. Assuming that his penalty kicks are independent, find the probability that Thanh first scores on his third shot.

Solution

This is a geometric distribution as:

> there are successes (scoring the penalty) and failures (missing)

> the trials are independent

> the probability of scoring is 0.3 and the probability of missing is 0.7.

The probability that he scores for the first time on his third shot is the probability that he misses on the first two shots and then scores.

P($X = 3$) = $0.3 \times 0.7^2 = 0.147$

Stop and think   Why might the geometric distribution not be a realistic model?

Example 9

The rules of a board game state that you need to roll a 6 on a fair die before you can start. Let R be the number of rolls up to and including the first 6. Find:

a   P($R = 4$)

b   P($R \leqslant 3$)

c   the probability that it takes more than four rolls to start.

Solution

a The probability of a success is $\frac{1}{6}$ so $R \sim \text{Geo}\left(\frac{1}{6}\right)$

$P(R = 4) = \frac{1}{6} \times \left(\frac{5}{6}\right)^3$

$= \frac{125}{1296}$

$= 0.0965$ (3 s.f.)

> $P(R = 4)$ is three failures and then a success.

b $P(R \le 3) = P(R = 1) + P(R = 2) + P(R = 3)$

$= \frac{1}{6} + \frac{1}{6} \times \frac{5}{6} + \frac{1}{6} \times \left(\frac{5}{6}\right)^2$

$= \frac{91}{216} = 0.421$ (3 s.f.)

c $P(R > 4) = P(R = 5) + P(R = 6) + P(R = 7) + P(R = 8) \ldots$
Since this carries on to infinity there needs to be another method to calculate this.

$P(R > 4) = 1 - P(R \le 4)$

> To have a success after the fourth roll, there must have been four failures.

Alternative solution 1:

$= 1 - \left(\frac{1}{6} + \frac{1}{6} \times \frac{5}{6} + \frac{1}{6} \times \left(\frac{5}{6}\right)^2 + \frac{1}{6} \times \left(\frac{5}{6}\right)^3\right)$

$= 1 - 0.5177\ldots$

$= 0.482$ (3 s.f.)

Alternative solution 2:

$P(R > 4) = \left(\frac{5}{6}\right)^4$

KEY INFORMATION
If $X \sim \text{Geo}(p)$, then $P(X > x) = (1 - p)^x$.

Exercise 3.4A

1 Find the probabilities below using your calculator, giving your answer to 3 d.p. where appropriate.

a For $X \sim \text{Geo}(0.2)$ find:

  i $P(X = 2)$ **ii** $P(X = 5)$ **iii** $P(X = 10)$.

b For $X \sim \text{Geo}(0.8)$ find:

  i $P(X = 2)$ **ii** $P(X = 5)$ **iii** $P(X < 3)$.

c For $X \sim \text{Geo}\left(\frac{1}{8}\right)$ find:

  i $P(X = 4)$ **ii** $P(X \le 4)$ **iii** $P(3 \le X < 7)$.

d For $X \sim \text{Geo}(0.4)$ find:

  i $P(X < 4)$ **ii** $P(X = 5)$ **iii** $P(X > 5)$.

e For $X \sim \text{Geo}(0.16)$ find:

  i $P(X \le 3)$ **ii** $P(X > 6)$ **iii** $P(X \ge 4)$.

2 A fair 10-sided die is rolled until a 7 is first seen. What is the probability of rolling the die six times?

3 On an island, the probability of rain each day is 0.5. Some children decide to play outside every day until it starts raining. Find the probability that the children play outside for 3 days.

4 A random variable X has the distribution Geo(p). Given that $p = 0.25$, find:

a $P(X < 5)$ **b** $P(X \geqslant 7)$.

5 A scientist is studying bream which are fish found in rivers. He knows that the probability of catching a bream is 35%. Find the probability that:

a he catches a bream for the first time on the sixth attempt

b it takes at least 10 attempts to catch a bream.

6 A factory engineer has to restart a furnace after it has been switched off. The probability that the furnace lights is 0.4. It takes 25 minutes to attempt to restart the furnace. Find the probability that she has restarted the furnace within 1 hour and 30 minutes.

7 Anya is collecting toys that are gifts in cereal packets. There are five toys in the set and the company put equal numbers of each toy in the packets. She has already collected four of the toys and wants the final toy to complete the selection. Find the probability that:

a she finds the missing toy in the fourth packet she opens

b she must buy more than five packets.

8 A company telephones its customers until they answer. On each call the probability that the customer answers the phone is s.

a What assumptions are needed to use the geometric distribution?

b Assuming that the assumptions are true, find the probability that the customer answers the phone on the nth try.

c The company has found that the probability that the person answers the phone on the second attempt is 0.21. It is known that the probability of a person answering the phone on the first attempt is more than 0.5. Find the probability of the customer answering the phone on the first phone call made.

9 A biologist is collecting data about beetles. She knows that one beetle in 12 has an orange body, and this occurs at random in the population. She needs to collect one beetle with an orange body.

a Find the probability that she needs to collect at least 20 beetles.

b Find the number of beetles she must collect to be 90% sure of obtaining one with an orange body.

10 Celine is counting cars going past the front gate of the school. She has been told that, on average, one car in 20 on the roads is yellow. Assuming that car colours are independent of each other, find the probability that:

a the first yellow car she sees is the 10th that passes

b the first yellow car is not among the first 25 cars

c there is at least one yellow car among the first 15 cars.

The binomial distribution

The **binomial distribution** is a discrete probability distribution that describes discrete random variables. As with the geometric distribution, the binomial distribution is concerned with successes and failures, but for a fixed number of trials.

In **Chapter 2 Probability, permutations and combinations** you learned about arrangements of various objects using $n!$ ('n factorial'). The **binomial coefficient** tells you how many ways there are of choosing unordered outcomes from all of the possibilities.

> If there are r events of one type, $n - r$ events of another type can be arranged in $\frac{n!}{r!(n - r)!}$ written as $\binom{n}{r}$ different orders.

If you were to flip a coin twice and wanted to know the probability of getting exactly one head, success would be heads and failure would be tails. The possible outcomes are HH, HT, TH or TT. However, you need to look at the number of arrangements for the heads. You could get a head and then a tail or a tail and then a head. You have two possible arrangements of getting one head out of four possible outcomes. The probability of flipping one head would be $\frac{1}{2}$.

If you were to flip a coin three times and wanted to know the probability of getting one head, success would be heads and failure would be tails. You could flip HHH, HHT, HTH, THH, HTT, THT, TTH, TTT. However, you need to look at the number of arrangements for the heads. You could get HTT, TTH or THT, so there are three possible arrangements of getting one head, out of eight possible outcomes. The probability for one head would be $\frac{3}{8}$.

Stop and think   We often use the binomial distribution to model coin tosses with the two outcomes as heads or tails. However, there is a chance with thicker coins that the coin lands on its edge. Why might the binomial distribution still be a suitable model?

It is important that you can identify situations that can be modelled using the binomial distribution. You can use the binomial distribution:

> if there are a fixed number of n independent trials

> if there are just two possible outcomes to each trial, success and failure, with fixed probabilities of p and q, respectively, where $q = 1 - p$

> if all trials are independent of each other

> if the probability of success, p, is the same in each trial.

The discrete random variable X is the number of successes in the n trials. X is modelled by the binomial distribution $B(n, p)$.

KEY INFORMATION
You can write $X \sim B(n, p)$.

The probability of success in a single trial is usually taken as p and the probability of failure as q. The probability of achieving exactly r successes in n trials is shown in this formula:

$$P(r \text{ successes in } n \text{ trials}) = \binom{n}{r} \times p^r \times (1-p)^{n-r}$$

An alternative way of writing this formula is P(r successes in n trials) = $^nC_r \times p^r \times q^{n-r}$ where $q = 1 - p$.

If a coin was flipped five times and you wanted to know the probability of getting three heads, you would need to use binomial distribution, as a written method would be time-consuming and cumbersome.

A success is heads and a failure is tails.

The number of arrangements of three heads in five coin flips is $^5C_3 = 10$.

$^5C_3$ is an alternative way of writing $\binom{5}{3}$. In general:
$$^5C_3 = \binom{n}{r} = \frac{n!}{r!(n-r)!}$$

The probability of success is $\frac{1}{2}$.

The probability of failure is $\frac{1}{2}$.

$P(3 \text{ successes}) = \frac{1}{2} \times \frac{1}{2} \times \frac{1}{2} = \left(\frac{1}{2}\right)^3$

$P(2 \text{ failures}) = \frac{1}{2} \times \frac{1}{2} = \left(\frac{1}{2}\right)^2$

Combining all of this gives $10 \times \frac{1}{8} \times \frac{1}{4} = 0.3125$.

Example 10

A coin is tossed 15 times. Find the probability of getting exactly five heads.

Solution

The probability of getting a head is $\frac{1}{2}$ and the probability of getting a tail is $\frac{1}{2}$. Therefore, the probability of getting five heads is $\left(\frac{1}{2}\right)^5$.

The total number of ways of tossing five heads in 15 trials is $\binom{15}{5}$, which gives 3003 arrangements.

$P(5 \text{ heads}) = \left(\frac{1}{2}\right)^5$

$P(10 \text{ tails}) = \left(\frac{1}{2}\right)^{10}$

Hence, the probability of throwing five heads in 15 trials is:
$$\binom{15}{5} \times \left(\frac{1}{2}\right)^5 \times \left(\frac{1}{2}\right)^{10} = 0.0916 \text{ (3 s.f.)}$$

This can also be displayed as $X \sim B(15, 0.5)$.

KEY INFORMATION

The probability that $X = r$ is given by:
$$P_r = \binom{n}{r} p^r (1-p)^{n-r}$$
n = number of trials
r = number of successes
$n - r$ = number of failures
p = probability of success in one trial

$\binom{15}{5}$ can be calculated using the ^nC_r button on your calculator.

Example 11

You are taking a multiple choice test containing 10 questions. If each question has four choices and you guess the answer to each question, what is the probability of getting exactly seven questions correct?

Solution

This is a binomial distribution because outcomes are either a success or a failure and are independent of each other. Therefore $X \sim B(10, 0.25)$. Using the formula:

$$P(\text{exactly 7 correct guesses}) = \binom{10}{7} (0.25)^7 (1 - 0.25)^3$$

$\approx 0.003\,09$ (3 s.f.)

Example 12

A particular set of traffic lights is on green 60% of the time. X is the number of drivers who pass through the traffic lights out of 30. What is the probability that exactly 11 drivers pass through?

Solution

Each driver represents a trial, so $n = 30$. The probability of success is 60%, so $p = 0.6$.

$X \sim B(30, 0.6)$

$$P(X = 11) = \binom{30}{11} \times 0.6^{11} \times (1 - 0.6)^{19} = 0.005\,45 \text{ (3 s.f.)}$$

In some situations, it is useful to be able to calculate probabilities across a range of different outcomes. For example, if $X \sim B(n, p)$ and you wanted to calculate $P(X \leq r)$, one method would be to say that:

$$P(X \leq r) = P(X = 0) + P(X = 1) + P(X = 2) + P(X = 3) + \cdots + P(X = r)$$

Depending on the values of n and r, it may sometimes be simpler to find the complement:

$P(X \leq r) = 1 - P(X > r)$

$$= 1 - (P(X = r + 1) + P(X = r + 2) + P(X = r + 3) + \cdots + P(X = n))$$

Example 13

$X \sim B(11, 0.41)$

Find the following probabilities:

a $P(X = 1)$

b $P(X = 0)$

c $P(X \geqslant 2)$.

Solution

$X \sim B(11, 0.41)$. So, $n = 11$, $p = 0.41$, $q = 0.59$

a $P(X = 1) = \begin{pmatrix} 11 \\ 1 \end{pmatrix} \times 0.41 \times 0.59^{10} = 0.0231$ (3 s.f.)

b $P(X = 0) = \begin{pmatrix} 11 \\ 0 \end{pmatrix} \times 0.41^{0} \times 0.59^{11} = 0.003\,02$ (3 s.f.)

c $P(X \geqslant 2) = 1 - (P(X = 0) + P(X = 1))$

$\qquad = 1 - (0.0231 + 0.003\,02)$

$\qquad = 0.974$ (3 s.f.)

> $P(X \geqslant 2) = 1 - P(X \leqslant 1)$

Example 14

A game of chance has probability of winning 0.73 and losing 0.27. Find the probability of winning more than seven out of 10 games.

> 'More than 7 games' means winning 8, 9 or 10 games.

Solution

The number of successes is a random variable $X \sim B(10, 0.73)$, assuming independence of trials.

$P(X > 7) = P(X = 8) + P(X = 9) + P(X = 10)$

$\qquad = \begin{pmatrix} 10 \\ 8 \end{pmatrix} \times 0.73^{8} \times 0.27^{2} + \begin{pmatrix} 10 \\ 9 \end{pmatrix} \times 0.73^{9}$

$\qquad \times 0.27^{1} + \begin{pmatrix} 10 \\ 10 \end{pmatrix} \times 0.73^{10}$

$\qquad = 0.466\,48...$

$\qquad = 0.466$ (3 s.f.)

So P(more than 7 wins in 10 games) $= 0.466$.

Stop and think   Could you use the complement of an event to solve **Example 14**?

In some situations, it is useful to be able to calculate probabilities across a specific range of outcomes. For example, if $X \sim B(n, p)$ and you want to calculate $P(r < X \leqslant s)$, you would say that:

> Make sure you check whether the inequalities include ($\leqslant$) or exclude ($<$) the values at either end of the range.

$$P(r < X \leqslant s) = P(X = r + 1) + P(X = r + 2) + \cdots + P(X = s)$$

Example 15

For $X \sim B(24, 0.84)$, find the probability that $19 < X \le 21$.

Solution

$19 < X \le 21$ shows that you need P($X = 20$ or 21)

$= P(X = 20) + P(X = 21)$

$= \binom{24}{20} \times 0.84^{20} \times (1 - 0.84)^4 + \binom{24}{21} \times 0.84^{21} \times (1 - 0.84)^3$

$= 0.426$ (3 s.f.)

Exercise 3.4B

1 Find the probabilities below using your calculator, giving your answer to 3 s.f. where appropriate.

a For $X \sim B(10, 0.2)$ find:

   i P($X = 2$)   **ii** P($X = 5$)   **iii** P($X = 10$).

b For $X \sim B(7, 0.8)$ find:

   i P($X = 2$)   **ii** P($X = 5$)   **iii** P($X < 3$).

c For $X \sim B\left(20, \frac{1}{8}\right)$ find:

   i P($X = 18$)   **ii** P($X \le 3$)   **iii** P($3 < X \le 18$).

d For $X \sim B(5, 0.4)$ find:

   i P($X < 4$)   **ii** P($X = 5$)   **iii** P($0 \le X < 3$).

e For $X \sim B(17, 0.16)$ find:

   i P($X \le 2$)   **ii** P($5 \le X < 8$)   **iii** P($2 < X \le 15$).

2 A fair six-sided die is rolled eight times. What is the probability of rolling exactly three sixes?

3 In basketball, shots scored from outside of the three-point line earn three points, shots from inside the three-point line are worth two points and a free throw from the free-throw line is worth one point. The probability of scoring is s. Each shot scored is independent of another. Write the probability formula to represent scoring six times from 10 shots.

4 A random variable X has the distribution B(12, p). Given that $p = 0.25$, find:

a P($X < 3$)   **b** P($X \ge 9$).

5 In a maths class, an average of three out of every five students ask for paper to write on. A random sample of 30 students is selected.

Find the probability that:

a exactly 10 students ask for paper

b fewer than 25 students ask for paper.

6 An airport has poor visibility 25% of the time in winter. A pilot flies to the airport 10 times during winter.

   a What is the probability that she encounters poor visibility exactly three times?

   b What is the probability that she encounters poor visibility at least three times?

7 There are 20 students in a maths class.

   a Find the probability that exactly one student has a birthday in January.

   b Find the probability that at most, four students have a birthday in January.

8 In a restaurant, 40% of the customers are in the age group 18–25.

   a Five customers are chosen at random. What is the probability that fewer than two of them are in the 18–25 age group?

   b If the five customers were chosen by picking them from the same table, justify the validity of applying the binomial distribution model.

9 A supermarket sells gingerbread men biscuits, and it is known that on average 3% of the gingerbread men biscuits are broken. A packet contains 10 gingerbread men biscuits. Assuming that the packet forms a random sample of the biscuits sold by the supermarket, calculate the probability that exactly two of the gingerbread men biscuits are broken.

10 Duck eggs produced at a farm are packaged in boxes of 10. Assume that, for any duck egg, the probability that it is broken when it reaches the retail outlet is 0.1, independent of all other eggs. A box is repacked if it contains two or more broken duck eggs. Calculate the probability that a randomly selected box needs to be repacked.

Mathematics in life and work: Group discussion

A football manager is interested to know whether one of his players is underperforming in league matches. Having observed the player in practice over a period of time, the manager knows that when the player has a shot on target at the goal, he scores 85% of the time. Having observed the player during recent league matches, the manager sees that the player has 12 shots on target at the goal and scores eight of them.

1 For some data sets it can be shown that around 95% of observed data will be within two standard deviations of the mean. Given this knowledge, what range of goals might reasonably be expected from the player from his 12 shots on target?

2 Considering your answer to **Question 1**, would you say that the player is underperforming? What other factors might you need to consider in making this kind of judgement? Are they all statistical?

3 The manager has a chance to use an alternative player for future matches. Having observed the alternative player in practice sessions over a period of time, the manager knows that when he has a shot on target at the goal, he scores 57% of the time. However, this player has recently had eight shots on target at the goal during league matches and has scored seven of them. The manager's own player has also scored seven goals during the same number of league matches, but which player would you choose for the next match and why?

3.5 Expectation and variance of geometric and binomial distributions

In **Sections 3.2 and 3.3** you saw that it was possible to find the expectation and variance of discrete random variables. In this section you will learn how to find the expectation of a geometric distribution and the expectation and variance of a binomial distribution.

Expectation of a geometric distribution

If the probability of rolling a 4 on a fair die is $\frac{1}{6}$, then it is expected that you would have to, on average, roll the die six times before you get a 4. This can be modelled by a geometric distribution, $X \sim$ Geo $\left(\frac{1}{6}\right)$ and E(X) = 6.

> In general for a geometric distribution E(X) = $\frac{1}{p}$.

Example 16

A reporter decides to ask people in the shopping centre if they watched the last football match, and interview the first person who watched. She knows that 5% of the town's population watched the football match. How many people should she expect to ask before she finds someone who watched the match?

Solution

Since we are asking people up to and including the first person who watched the match, we can use a geometric distribution, $X \sim$ Geo(0.05).

$$\text{E}(X) = \frac{1}{p}$$

$$= \frac{1}{0.05}$$

$$= 20 \text{ people}$$

Stop and think   Suppose the reporter had asked 35 people and none of them had watched the football game. Would this suggest that the 5% figure might be inaccurate or might it just have happened by chance?

Expectation and variance of a binomial distribution

A supermarket sells eggs in boxes of 12. The probability of an egg being cracked is $\frac{1}{6}$. We would expect that there would be two broken eggs per box since one out of every six eggs is broken and there are 2 eggs. Hence $\frac{1}{6} \times 12 = 2$.

In this case, X is the number of broken eggs in a box and can be modelled by $X \sim$ B$\left(12, \frac{1}{6}\right)$.

> Remember that Var(X) can also be written as σ^2.

> **KEY INFORMATION**
> In general for a binomial distribution, E(X) = $\mu = np$.

If X and Y are two independent random variables, then:

$$E(X + Y) = E(X) + E(Y)$$

$$\text{Var}(X + Y) = \text{Var}(X) + \text{Var}(Y)$$

Extending these results for multiple independent random variables X_1, X_2, etc., we can say that:

$$E(X_1 + X_2 + X_3 + \ldots + X_n) = E(X_1) + E(X_2) + E(X_3) + \ldots + E(X_n)$$

$$\text{Var}(X_1 + X_2 + X_3 + \ldots + X_n) = \text{Var}(X_1) + \text{Var}(X_2) + \text{Var}(X_3) + \ldots + \text{Var}(X_n)$$

If we consider the random variable $X \sim B(n, p)$ and the random variables Y_i, which represent the number of successes on the ith trial ($i \leqslant n$), then the distribution of each Y_i will be:

y	0	1
$P(Y = y)$	$1 - p$	p

$$E(Y_i) = 0(1 - p) + (1 \times p) = p$$

$$\text{Var}(Y_i) = (1^2 - p) - p^2 = p(1 - p)$$

Since $X = Y_1 + Y_2 + Y_3 + \ldots + Y_n$

$$E(X) = E(Y_1) + E(Y_2) + E(Y_3) + \ldots + E(Y_n)$$

$$= p + p + p + \ldots + p$$

$$= np$$

$$\text{Var}(X) = \text{Var}(Y_1) + \text{Var}(Y_2) + \text{Var}(Y_3) + \ldots + \text{Var}(Y_n)$$

$$= p(1 - p) + p(1 - p) + p(1 - p) + \ldots + p(1 - p)$$

$$= np(1 - p) \qquad or \qquad \text{Var}(X) = npq$$

You can find the standard deviation of a binomial distribution by taking the square root of the variance, which gives standard deviation, $\sigma = \sqrt{np(1 - p)}$.

KEY INFORMATION

If $X \sim B(n, p)$, then the variance and standard deviation are given by:
$\sigma^2 = np(1 - p)$ and
$\sigma = \sqrt{np(1 - p)}$

Example 17

After being given a particular type of vaccine, mild side effects can appear in 10% of patients. A doctor's surgery gives this vaccine to 50 babies. If X is the random variable representing the number of patients who display the side effects, find an appropriate distribution to model X, the expectation of X and its variance.

Solution

There is a fixed number of trials ($n = 50$) and a fixed probability of side effects ($p = 0.1$). Therefore a binomial distribution is appropriate.

So, $X \sim B(50, 0.1)$.

The expected number of babies who display side effects is $E(X) = np = 50 \times 0.1 = 5$.

The variance of X is σ^2, $np(1 - p) = 50 \times 0.1 \times 0.9 = 4.5$.

Example 18

A biased coin is flipped 200 times and the coin shows tails 140 times. The coin is then flipped 15 more times. Find:

a the probability that you will get exactly 10 tails

b the expected number of tails

c the standard deviation of the number of tails

d $P(X > \mu + \sigma)$.

Solution

a The probability of getting tails $= \dfrac{140}{200} = 0.7$.

> The coin is biased so we have to use the experimental probability.

X is the number of tails so $X \sim B(15, 0.7)$.

$$P(X = 10) = \binom{15}{10} \times 0.7^{10} \times 0.3^5 = 0.206 \text{ (3 d.p.)}$$

b $E(X) = n \times p = 15 \times 0.7 = 10.5$

> Look out for this when interpreting the numerical answer for the expected value. Some answers will not make sense unless they are rounded to an integer.

The expected number of tails is 11, as you can't have half a tail from flipping a coin. The answer is rounded up.

c $\sigma^2 = np(1 - p) = 15 \times 0.7 \times 0.3 = 3.15$

$\sigma = \sqrt{3.15} = 1.775$ (3 d.p.)

> Remember that μ is another notation for $E(X)$.

d $P(X > \mu + \sigma) = P(X > 10.5 + 1.775)$

$= P(X > 12.275)$

$= P(X = 13) + P(X = 14) + P(X = 15)$

> Since the binomial distribution is discrete you need to take the integers that are greater than 12.275.

$$= \binom{15}{13} \times 0.7^{13} \times 0.3^2 + \binom{15}{14} \times 0.7^{14} \times 0.3^1$$

$$+ \binom{15}{15} \times 0.7^{15}$$

$= 0.127$ (3 s.f.)

Exercise 3.5A

1 Find E(X) for each of the following distributions:

    a $X \sim \text{Geo}(0.3)$

    b $X \sim \text{Geo}(0.45)$

    c $X \sim \text{Geo}\left(\dfrac{3}{7}\right)$.

2 For each of the distributions below, find:

    i μ **ii** σ^2.

    a $X \sim \text{B}(10, 0.2)$

    b $X \sim \text{B}(150, 0.3)$

    c $X \sim \text{B}\left(20, \dfrac{3}{8}\right)$

    d $X \sim \text{B}(5, 0.1)$.

3 A fair die is rolled.

    a Find the expected number of rolls before it lands on a 4.

    b Find the expected number of 4s seen if the die is rolled 42 times.

4 X is a random variable where $X \sim \text{B}(30, 0.1)$. Find:

    a the expectation and variance

    b $P(\mu \leqslant X < \mu + \sigma)$.

5 Find the expectation and standard deviation of the random variable $X \sim \text{B}(25, 0.6)$.

    Hence find the probability the random variable takes a value within one standard deviation of the expectation.

6 The random variable $X \sim \text{B}(n, p)$ has an expectation of 2.5 and a variance of 1.875.

    Find the values of n and p.

(MM) 7 It is estimated that 13 of every 20 people wear glasses. A researcher wants the expected number of people who wear glasses in her sample to be 39.

    a How many people does she need to select for her sample?

    b What is the probability that there are exactly 39 people who wear glasses in her sample?

(PS) (MM) 8 A factory makes plates and sells them in boxes of eight. The probability that a plate is damaged is 0.1.

    a How many plates would the factory expect to be damaged in each box?

    b What is the probability of one or more damaged plates in a box?

    After packing the plates into boxes of eight, the factory then sells them to shops in batches of 40 boxes.

   c Find the expected number of boxes that contain one or more damaged plates in a batch.

   d Explain why the expected number of boxes found in part **c** needs to be rounded to the nearest integer.

9 A bag contains five $1 coins and 10 similar foreign coins. Jack selects one coin at random from the bag. He keeps the coin if it is a $1 coin, otherwise he returns it to the bag and selects again.

   a Find the expected number of selections that Jack makes, up to and including the first $1 coin.

   b Given that Jack picks a $1 coin on his first selection, find the expected number of extra selections that Jack makes to obtain a second $1 coin.

10 A chocolate egg cost $1.20. One chocolate egg in 12, on average, contains a limited edition toy. If a child buys a chocolate egg every week, find the expected total cost of the chocolate eggs bought, up to and including the first egg with a limited edition toy.

SUMMARY OF KEY POINTS

> A discrete random variable is a random variable that can only take set values.

> All the probabilities in a probability distribution must add up to 1.

> You can find the expectation of a discrete random variable using $E(X) = \mu = \sum xp$.

> You can find the variance of a discrete random variable using $Var(X) = \sigma^2 = \sum (x - \mu)^2 p = \sum x^2 p - \mu^2$.

> The geometric distribution is a discrete probability distribution that describes discrete random variables up to and including the first success. It is written as $X \sim Geo(p)$.

> The expectation of the geometric distribution is $E(X) = \mu = \dfrac{1}{p}$.

> You can use the geometric distribution:

> > if there are just two possible outcomes to each trial, success and failure, with fixed probabilities of p and q, respectively, where $q = 1 - p$

> > if all trials are independent of each other

> > if the probability of success, p, is the same in each trial.

> The binomial distribution is a discrete probability distribution that describes discrete random variables for a fixed number of trials. It is written as $X \sim B(n, p)$.

> The expectation of the binomial distribution is $E(X) = \mu = np$ and the variance is $Var(X) = \sigma^2 = np(1 - p)$.

> You can use the binomial distribution:

> > if there are a fixed number of n trials

> > if there are just two possible outcomes to each trial, success and failure, with fixed probabilities of p and q, respectively, where $q = 1 - p$

> > if all trials are independent of each other

> > if the probability of success, p, is the same in each trial.

EXAM-STYLE QUESTIONS

1 Part of the probability distribution of a random variable X is shown in the table below.

x	0	1	2	3
$P(X = x)$	0.3		0.4	0.05

   a Find $P(X = 1)$.

   b Find $E(X)$.

2 The random variable $T \sim B(20, 0.4)$. Find:

   a $P(T = 3)$

   b $P(T \leqslant 3)$

   c $E(T)$ and $Var(T)$.

Ⓒ **3** Amelie is trying to light a match. The number of attempts she makes up to and including her first success is a random variable denoted by X.

   a State two assumptions that are needed in this context for a geometric distribution.

   Given that X is $Geo(0.4)$, calculate:

   b the probability that she lights the match on the third attempt

   c the probability that she will take longer than three attempts.

4 There are four cards with numbers on them as shown below.

| 1 | 3 | 0 | 3 |

   S is the random variable that denotes the sum of two cards drawn without replacement.

   a Find $P(S = 4)$.

   b Find the probability distribution table for S.

   c Calculate the expectation and variance of S.

5 Kartik and Khari decide to have a competition in the common room, playing each other at pool. They decide to play 10 frames of pool to see who is better.

   The probability that Kartik will win any frame is 0.21. The outcome of each frame is independent of the outcome of every other frame.

   Find the probability that during the ten frames, Kartik wins:

   a exactly two frames

   b at least four frames

   c more than four but fewer than nine frames.

ⓂⓂ **6** Pawel plants 15 randomly selected seeds in a planter. He knows that 10% of the seeds will yield red flowers while the others will yield white flowers. All the seeds grow successfully and Pawel counts the number of red flowers and the number of white flowers.

   a Find the probability of three flowers being red.

   b Find the probability of all the flowers being red.

   c Find the expected number of red flowers.

ⓂⓂ **7** Lucas is practising his basketball shots. His chance of scoring is 0.3 and each shot is independent.

   Find the probability that:

   a he first succeeds on his fourth throw

   b he first scores after his fourth throw

   c he scores for the second time on his fourth throw.

Stefan joins him and they take turns to try to score. Stefan scores once out of every five shots. Lucas goes first and they take turns until one of them scores.

d Find the probability that Lucas first scores on his second shot.

e Find the probability that Stefan first scores on his seventh attempt.

8 R is a discrete random variable with the probability distribution:

r	-2	3
$P(R = r)$	a	$1 - a$

a Show that $\text{Var}(R) = 25a(1 - a)$.

b Given that the $E(R) = 2$, find the standard deviation of R.

9 A factory that makes zips knows that the probability of a zip being faulty is 0.05. In a random sample of 30 zips, find the probability that:

a two zips are faulty

b more than three zips are faulty.

The factory sells the zips in packets of 30 zips. A clothes manufacturer buys 20 packets. Calculate:

c the probability that there are exactly five packets with more than three faulty zips

d the expected number of packets with more than three zips faulty.

10 Two parents each have a genetic medical condition and there is a $\frac{1}{7}$ chance that each of their children will develop the condition. Given that the parents have three boys and two girls, what is the probability that any three or any two of the children develop the condition?

11 A bag contains eight red counters, six black counters and four green counters. Six counters are drawn at random from the bag. Find the probability that:

a exactly four of the counters are red

b at least four of the counters are red.

12 The probability of the internet being offline during a visit to an internet café is 0.01. A sign offers a full refund if the internet is offline during one or more of your first 10 visits.

a What is the probability of getting a refund?

b What is the probability that exactly three out of five visitors receive a refund?

13 A jar contains 20 red beads, 30 yellow beads, 40 green beads and 50 blue beads. Ten beads are selected at random to form a necklace. What is the probability that the necklace contains exactly six beads of the same colour?

14 The probability of a hunter hitting a target with any shot is 0.7. Given that the hunter shoots six bullets, find the probability that he hit the target at least three times.

15 A spinner has three different colours, red, blue and yellow, with probability $\frac{1}{2}, \frac{1}{3}$ and $\frac{1}{6}$, respectively. The random variable X denotes the result of the combination of two consecutive outcomes.

   a State the assumptions made.

   b Construct a probability distribution table for X.

Two players, Adam and Bob, play a game with a biased spinner like the one described above and a fair spinner with one red, one blue and one yellow section. Adam chooses one of the spinners at random and spins it. If it lands on yellow, he wins a point.

   c Calculate the probability that Adam wins a point.

   d Bob chooses a spinner, spins it and wins a point. Find the probability that he has at least three attempts before he wins.

16 Some scientists suggest that there will be a decrease in rainfall in the Amazon. The probability that it rains in any given week during the 'dry' season (16 weeks between July and October) can be taken to be 0.35. Use a binomial distribution to calculate the probability of rainfall in more than three weeks of the 'dry' season. State two assumptions needed for a binomial distribution to be a good model.

17 A computer program generates a single positive digit at random.

   a Write down the proportion of the digits produced that you would expect to be sixes.

   b Calculate the probability that the first twelve digits produced contain no sixes.

After running the first 12 digits, the computer contentiously generates digits until a six is produced, and the number, N, of digits up to and including the first six is recorded.

   c Find the probability that $N = 1$.

   d Calculate the probability that N is exactly 9.

18 When a driving theory test is taken, the probability of passing is 0.65 at any attempt. If a candidate fails, tests continue until the candidate passes. The random variable X is the number of tests a candidate takes to pass the test.

   a State a distribution that can be used as a model for X.

   b Find the probability that a candidate passes the first time.

   c Find the probability that a candidate will need five or fewer attempts to pass the test.

   d In a sample of 200 candidates, find how many you would expect to pass at the first attempt.

19 On average, Keith scores one goal from a free kick in every five attempts.

   a Name the distribution that models the number of attempts needed to score his first goal.

   b Find the probability that he scores his first goal on his fifth attempt.

   c Find the probability that he scores his first goal after at least four failures.

   d What assumption was made?

20 A machine cuts plastic to given measurements. Normally only 4% of the pieces are of unacceptable quality. Each plastic piece is inspected as it leaves the machine during a run. Let X denote the number of plastic pieces inspected up to and including the first unacceptable piece. If $X \leqslant 3$, then the machine will be stopped and serviced.

   a Calculate the probability that the machine is stopped during a run.

   b Calculate $P(X < E(X))$.

   c In a set of 10 runs, find the expectation and variance of the number of times the machine is stopped. State any necessary assumption for the validity of your model.

Mathematics in life and work

In a factory, a machine is used to put an exact number of bolts into individual bags to be sold in hardware shops. The manufacturers of the machine say that the machine is 99.9% accurate – that is, it should put the correct number of bolts into the bag on $\frac{999}{1000}$ occasions.

1 What is the probability of the machine producing exactly 99 accurate bags out of a batch of 100 bags?

2 What is the probability of the machine producing at least 98 accurate bags out of a batch of 100 bags? Give your answer to 4 decimal places.

The manufacturer has agreed to service the machine if there are more than three inaccurate bags in any batch of 100.

3 What is the probability that the manufacturer will need to service the machine?

The manufacturer has agreed to replace the machine if there are more than four inaccurate bags in any batch of 100.

4 What is the probability that the manufacturer will need to replace the machine?

4 NORMAL DISTRIBUTION

Mathematics in life and work

In this chapter you will learn how to use the normal distribution to calculate missing information, such as the mean and standard deviation, and to find the probability of an event occurring. The modelling of situations as a normal distribution has a wide range of applications in many different careers. For example:

➤ If you were an examiner and you wanted to set the grade boundaries of a test so that the top 10% of students gained an A*, you could model the scores of students as a normal distribution.

➤ If you were a doctor, you could model the weights of babies as a normal distribution. This would allow you to work out at which percentile of the population a baby's weight lies and plan treatment accordingly.

➤ If you were an engineer in a canning factory and you knew that a machine has a set standard deviation, you could work out the settings for the machine so that 90% of the cans contain more than a minimum volume.

This chapter includes some of the ways you could use the normal distribution if you were an engineer.

LEARNING OBJECTIVES

You will learn how to:

➤ understand and use the normal distribution to model continuous random variables

➤ analyse the shape and symmetry of the normal distribution

➤ find probabilities using the normal distribution table, given the values of μ and σ

➤ find μ and σ given probabilities

➤ link the normal distribution to the binomial distribution

➤ apply continuity corrections.

LANGUAGE OF MATHEMATICS

Key words and phrases you will meet in this chapter

➤ binomial distribution, binomials, continuity correction, normal distribution, standardised normal distribution, standardising

PREREQUISITE KNOWLEDGE

You should already know how to:

› interpret the mean and standard deviation of a data set

› draw histograms

› infer properties of populations or distributions from a sample, whilst knowing the limitations of sampling

› apply statistics to describe a population.

You should be able to complete the following questions correctly:

1 The number of calls received by the emergency services in 20 consecutive minutes was:
6, 1, 3, 4, 0, 4, 2, 3, 9, 4, 4, 3, 5, 0, 4, 6, 6, 4, 2, 5.

Calculate:

a the mean

b the standard deviation.

In the next five minutes the following numbers of calls were made: 0, 0, 2, 5, 1.

c Calculate the new mean and standard deviation.

2 The random variable Y has the following distribution:

Y	1	3	4	6
P($Y = y$)	$\frac{1}{12}$	$\frac{5}{12}$	$\frac{1}{3}$	$\frac{1}{6}$

a Find the mean and standard deviation of Y.

b Find $P(Y > \mu + \sigma)$.

4.1 The normal distribution

If you measured the heights of all the people at a concert on any particular night and plotted the results, the graph would look something like the histogram on the right.

By plotting the continuous variable of height, you can represent all the possible probabilities of any person being any height. Continuous distributions are distributions for continuous variables and are more commonly known as probability densities. A bell-shaped probability density like this is called the **normal distribution**.

Each end of the bell curve extends asymptotically. The y-axis in the normal distribution represents the probability density. For example, in the diagram, the total area under the curve represents 1.

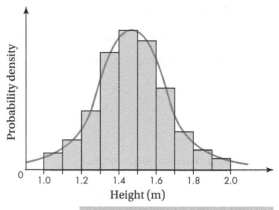

'Asymptotically' means that the curve never touches the axis.

KEY INFORMATION

Normal distributions deal with continuous variables.

Here are some features of a curve that describes a continuous distribution like the normal distribution:

> The area under the curve equals 1.

Stop and think | Can you explain why the area under the curve equals 1?

KEY INFORMATION
The area under the curve equals 1.

Using the example of heights at a concert, you would conclude that the probability that a particular person has a height is 1.

> The probability of any exact value of X is 0.

In the concert example, the probability that a particular person will have a height of exactly 1.3 m is essentially zero.

KEY INFORMATION
The probability of any exact value of X is 0.

> The area under the curve and bounded between two given points on the x-axis is the probability that a number chosen at random will fall between the two points.

In the concert example, suppose that the probability of measuring between 1.3 m and 1.4 m is $\frac{1}{10}$. Then the continuous distribution for possible measurements would have a shape that places 10% of the area below the curve in the region bounded by 1.3 m and 1.4 m on the x-axis.

Stop and think | What would the histograms look like if you had separated the heights of males and females?

> Normal distributions are defined by two parameters, the mean (μ) and the standard deviation (σ).

> 68% of the area of a normal distribution is within one standard deviation of the mean.

> Approximately 95% of the area of a normal distribution is within two standard deviations of the mean.

> 99.75% of the data lies within ±3 standard deviations of the mean.

> The mean, μ, and standard deviation, σ, determine the shape of the normal curve.

> The mean gives the central location of the data, which is the line of symmetry.

> The smaller the standard deviation, the less spread out the data. The larger the standard deviation, the more spread out the data.

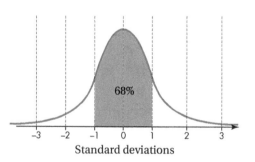

Standard deviations

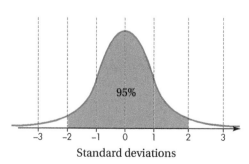

Standard deviations

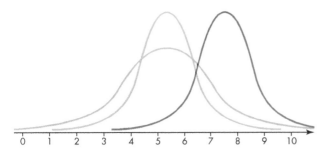

In the diagram above, the red and green curves have the same shape and width. This means that the standard deviations are the same. The green and blue curves have the same mean. The blue curve is wider, so it has the larger standard deviation, because its data is more spread out.

A point of inflexion is where the point on the curve stops increasing and starts decreasing.

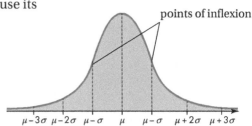

points of inflexion

$\mu-3\sigma$ $\mu-2\sigma$ $\mu-\sigma$ μ $\mu-\sigma$ $\mu+2\sigma$ $\mu+3\sigma$

> **Stop and think** Why is the peak of the curve not a point of inflexion?

The inflexion points of the normal distribution curve are located at $\mu - \sigma$ and $\mu + \sigma$.

Example 1

The marks in an examination are normally distributed.
a What percentage of the marks are above the mean score?

b What percentage of the marks are between −1 and +1 standard deviation?

c What percentage of the marks are between the mean and +2 standard deviations?

d What percentage of the marks would be between −2 standard deviations and +1 standard deviation?

Solution

a As the normal distribution is symmetrical on either side of the mean, 50% of the marks will be above the mean.

b You would expect 68% of the marks to be between −1 and +1 standard deviations of the mean.

c 47.5% of the marks will be between the mean and +2 standard deviations because 95% of the data is between −2 and +2 standard deviations of the mean.

In other words, you would expect 16% of the marks on the left side of −1 or right side of +1 standard deviation.

It is known that 95% of the data is between −2 and +2 standard deviations on either side of the mean. Therefore, 2.5% of the data is on the left side of −2 standard devision below the mean. There is also 2.5% of the data on the right side of +2 standard devision.

d Between −2 and +2 standard deviations is 95% of the data and between −1 and +1 standard deviations is 68% of the data.

So from −2 standard deviations to the mean is 47.5% of the data, and from the mean to +1 standard deviation is 34% of the data. Combining these gives 81.5% of the data.

Exercise 4.1A

1 Correct the following list of properties of a normal distribution.

    a The distribution is symmetrical about the standard deviation.

    b The mode, median and mean are all different.

    c The total area above the curve is 1.

    d The distribution is defined by three parameters: the mean, the median and the standard deviation.

(c) 2 Comment on the features of the two distributions in the diagram.

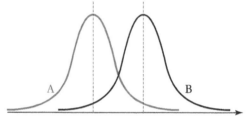

(PS) 3 The heights of some Year 13 students are normally distributed.

    a What percentage of the Year 13 students are above the mean height?

    b What fraction of the Year 13 students have heights between the mean and +1 standard deviation?

    c What percentage of the Year 13 students have heights between the mean and +2 standard deviations?

    d What fraction of the Year 13 students have heights between −3 standard deviations and +2 standard deviations?

(c) 4 Compare the curves A and B in the diagram on the right. Describe the curves using the terms 'mean' and 'standard deviation'.

5 A continuous random variable, X, is normally distributed with mean 42.5 and standard deviation 1.5.

Another continuous random variable, Y, is also normally distributed with mean 42.5 and standard deviation 2.5.

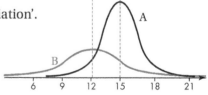

Draw the probability density curves of the random variables X and Y on the same graph.

6 a Use the following statements to identify each distribution graph.

    Statement A: The random variable A has the smallest standard deviation.

(c) **Communication**   (MM) **Mathematical modelling**   (PS) **Problem solving**

Statement B: The random variable B has the lowest mean.

Statement C: The random variable C has mean 7.2 and standard deviation 1.2.

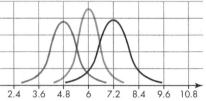

b State the mean of random variable B.

7 Which of the following statements are **i** always true, **ii** sometimes true, **iii** never true?

a The line of symmetry of the normal distribution is located at the mean, μ, of the distribution.

b There is a non-stationary point of inflexion on each side of the mean on a normal distribution curve. This is located two standard deviations from the mean.

c A normal distribution is a continuous distribution, so it is not possible to find the probability that a random variable takes a specific value.

d The normal distributions have a distinctive bell-shaped curve, symmetrical, with scores more concentrated in the middle than in the tails.

8 A manufacturer produces screws, the diameter of which can be modelled by the normal distribution. Its graph is shown on the right.

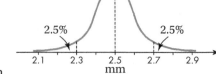

a State the mean and standard deviation of the screws.

b Find the percentage of screws that have a diameter measurement within one standard deviation of the mean.

9 Assume that the height of adult women can be modelled by the normal distribution with mean 167 cm and standard deviation 1.8 cm.

a Sketch a normal distribution curve.

b Find the probability that an adult woman, chosen at random, will have a height greater than 169.7 cm.

10 The results of an examination can be modelled by a normal distribution with mean 68 and standard deviation 15. Only 16% of the students failed the exam. Find the pass mark.

> **Stop and think** Can you think of other situations that might produce a symmetrical distribution like a normal distribution?

4.2 Using the normal distribution

You can calculate the area under each section of a normal distribution curve using integration.

The diagram on the right shows a normal distribution with a mean of 24 and a standard deviation of 1. The shaded area between 23 and 25 contains 68% of the

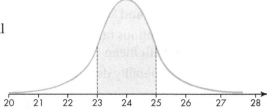

distribution – so the probability of getting a value between 23 and 25 would be 0.68. Similarly, 95% of the data lies between 22 and 26 and anything below 21 or above 27 would be classed as an outlier.

The area under a normal curve shows probabilities between two values. The probability of the variable taking a value between two limits is the area under the curve between those limits. Not all probabilities will be as straightforward as 1 or 2 standard deviations, so you will need to use a formula.

If a variable X has a normal distribution then you can write this using mathematical notation:

$$X \sim N(\mu, \sigma^2)$$

where:

> X is the random variable (this can be any letter)

> ~ is shorthand for 'is distributed'

> N tells you that it has a normal distribution

> μ is the mean

> σ is the standard deviation

> σ^2 is the square of the standard deviation and is called the variance.

For example, if $X \sim N(25, 1^2)$, this tells you that:

> the random variable X has a normal distribution

> with a mean $\mu = 25$

> and a standard deviation $\sigma = 1$.

This means that 68% of the values lie between 24 ($= 25 - 1$) and 26 ($= 25 + 1$).

Remember that the second parameter is the variance. The standard deviation is the square root of this number.

Example 2

The heights of males and females in a town were recorded in cm and the results were normal distributions as follows:

Males: $X \sim N(178, 7^2)$

Females: $Y \sim N(167, 64)$

Write down the mean, variance and standard deviation for each gender.

Solution

Using $X \sim N(\mu, \sigma^2)$, you know that the first parameter is the mean and the second is the variance. With that in mind:

Males: $X \sim N(178, 7^2)$ has a mean of 178 cm, variance of 49 cm (or 7^2) and standard deviation of 7 cm.

Females: $Y \sim N(167, 64)$ has a mean of 167 cm, variance of 64 cm (or 8^2) and standard deviation of 8 cm.

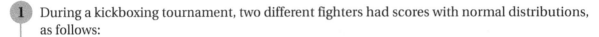

Exercise 4.2A

1 During a kickboxing tournament, two different fighters had scores with normal distributions, as follows:

Fighter A: mean = 3, variance = 4

Fighter B: mean = 2, standard deviation = 3

Use the correct mathematical notation to describe the distribution of each fighter's scores.

2 A school took part in a Maths Challenge over two different years. The scores have normal distributions as follows:

Year 1: $\mu = 0$, $\sigma^2 = 5$

Year 2: $\mu = -1$, $\sigma = 2$

Use the correct mathematical notation to describe the distribution of the scores for each year.

(C) 3 A mathematics student was asked to write down the mathematical notation for the scores on an app that had a normal distribution with mean $\mu = 7$ and standard deviation $\sigma = 11$.

The student wrote $N \sim X(11, 49)$.

Write down everything that is wrong with the student's notation, then write down the correct notation.

4 Two mathematics students sat a set of tests and had results with normal distributions, as follows:

Student A: $X \sim N(5, 7^2)$

Student B: $Y \sim N(6, 81)$

a Write down the mean, variance and standard deviation of the students' scores.

b Sketch the two distributions on one graph.

5 The random variables X and Y are normally distributed with the same mean. The variance of the random variable X is four times as large as that of the random variable Y. Plot both random variables on the same graph.

6 The random variables X and Y can be modelled by a normal distribution. $X \sim N(4.8, 2.25)$ and $Y \sim N(4.8, 1.5^2)$.

Adam makes the following statement:

The mean of the random variable X is the same as the mean of the random variable Y. The variance of the random variable X is greater than the variance of the random variable Y.

Billy disagrees, claiming that random variables X and Y have the same mean and the same variance, therefore the random variable X is the random variable Y.

Who is right: Adam, Billy or neither? State your reasons.

7 Some sketches of continuous distribution graphs are shown below.

a
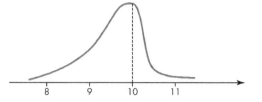

b
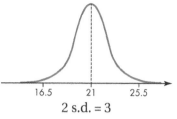

2 s.d. = 3

c
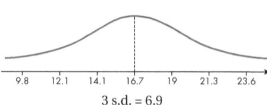

3 s.d. = 6.9

d
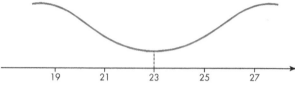

   i Giving reasons, identify which of these are normal distribution curves.

   ii Use the correct mathematical notation to describe the normal distribution curves.

8 State, with a reason, whether each of the following statements is always, sometimes or never true.

   a $X \sim N(4^2, t^2)$ describes a random variable X, which is normally distributed with mean 4^2 and standard deviation t^2.

   b Any continuous variable can be modelled by a normal distribution with mean μ and standard deviation σ.

9 The reaction times, in minutes, of some science experiments can be modelled by normal distributions, as follows:

Experiment A: $T_A \sim N(7, 0.5^2)$

Experiment B: $T_B \sim N(5.5, 0.2)$

Experiment C: $T_C \sim N(6.4, 0.3^2)$

   a Which experiment has the shortest average reaction time?

   b Yuki suggests that the Experiment B has the best outcome. Explain why her suggestion is correct.

10 The mass of mixed nuts packed by a machine is normally distributed with a mean of 50 grams and a standard deviation of 2.3 grams.

   a Use the correct mathematical notation to describe the distribution of the weights.

   b Sketch this as a graph.

The standardised normal distribution

It is not necessarily the case that normal distributions have the same means and standard deviations. A normal distribution with a mean of 0 and a standard deviation of 1 is called the **standardised normal distribution**.

The standardised normal distribution can be used to work out probabilities for variables with a normal distribution. The standardised normal distribution is defined as follows:

$$Z \sim N(0, 1)$$

This tells you that:

> the standardised normal random variable is denoted by Z

> Z has a normal distribution

> Z has mean $\mu = 0$

> Z has standard deviation $\sigma = 1$.

The diagram below shows the standardised normal distribution.

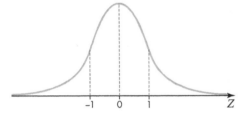

KEY INFORMATION
$Z \sim N(0, 1)$ is the standardised normal distribution.

An uppercase Z signifies the random variable that is being standardised. A lowercase z signifies the z-score.

Any normal distribution can be made to fit the standardised normal distribution. The process is called **standardising**. Since the distribution has a mean of 0 and a standard deviation of 1, the value of Z is equal to the number of standard deviations below (or above) the mean. This is often referred to as the z-score. For example, a z-score of -2.5 represents a value 2.5 standard deviations below the mean. Section 4.3 illustrates how to standardise a normally distributed continuous random variable.

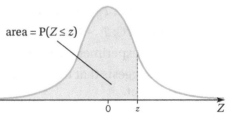

On the right is a graph of the standardised normal distribution. It is symmetrical about the mean, 0. The shaded area under the curve represents the probability that Z is less than or equal to z. To calculate this probability, you need to find the area, shaded in green.

The cumulative distribution function, $\Phi(z)$, finds the probability that $Z \leqslant z$. This corresponds to the area under the curve to the left of Z for different values of z.

Φ is the capital Greek letter F called phi, pronounced 'fi'.

There are three different cases to consider.

$P(Z < a) \equiv P(Z \leqslant a)$.

1 Z is less than value a: $(Z < a)$

2 Z is greater than value a: $(Z > a)$

KEY INFORMATION

3 Z is between values a and b: $(a < Z < b)$.

$\Phi(z) = P(Z \leqslant z)$

Stop and think   When calculating $\Phi(z)$, why can we say that $P(Z \leqslant a)$ is no different to $P(Z < a)$? Can we say the same about $P(Z \geqslant a)$ and $P(Z > a)$?

If you are given $Z \sim N(0, 1)$, you can say that the probability that Z is less than a value a (where $a > 0$) is shown by:

$P(Z < a)$

where P is the probability.

You know that the mean of the standard normal distribution is 0, so the diagram will look like this:

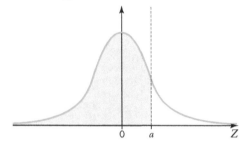

If you are given $Z \sim N(0, 1)$, you can say that the probability that Z is greater than a value a (where $a > 0$) is shown by:

$P(Z > a)$

where P is the probability.

You know that the mean of the standard normal distribution is 0, so the diagram will look like this:

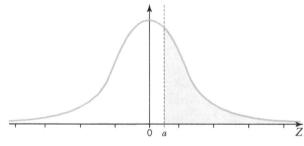

If you are given $Z \sim N(0, 1)$, you can say that the probability that Z is between the value a and b (where $a, b > 0$ and $b > a$) is shown by:

$P(a < Z < b)$

where P is the probability.

You know that the mean of the standard normal distribution is 0, so the diagram will look like this:

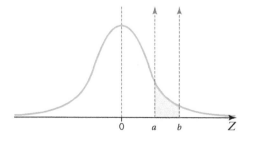

Using tables for the normal distribution

Cumulative probabilities associated with the z-score can be found in normal distribution tables. These give values for $P(Z < z)$, where $z > 0$.

If you are given $Z \sim N(0, 1)$ and are asked to find, for example, $P(Z < 0.432)$, begin by drawing a clear diagram.

The expression for the probability is:

$$P(Z < 0.432) = \Phi(0.432)$$

where Φ represents the area to the left of the value in brackets – that is, any given value of Z.

The normal distribution table splits the z-value into two sections: a two-decimal-place value and the thousandths digit. The first part of the table gives the probability and the thousandths digit table tells you what to add to the probability.

$\Phi(0.432) = 0.6664 + 0.0007$

$\qquad = 0.6671$

z	0	1	2	3	4	5	6	7	8	9	1	2	3	4	5	6	7	8	9
															ADD				
0.0	0.5000	0.5040	0.5080	0.5120	0.5160	0.5199	0.5239	0.5279	0.5319	0.5359	4	8	12	16	20	24	28	32	36
0.1	0.5398	0.5438	0.5478	0.5517	0.5557	0.5596	0.5636	0.5675	0.5714	0.5753	4	8	12	16	20	24	28	32	36
0.2	0.5793	0.5832	0.5871	0.5910	0.5948	0.5987	0.6026	0.6064	0.6103	0.6141	4	8	12	15	19	23	27	31	35
0.3	0.6179	0.6217	0.6255	0.6293	0.6331	0.6368	0.6406	0.6443	0.6480	0.6517	4	7	11	15	19	22	26	30	34
0.4	0.6554	0.6591	0.6628	0.6664	0.6700	0.6736	0.6772	0.6808	0.6844	0.6879	4	7	11	14	18	22	25	29	32

So $P(Z < 0.432) = 0.6664 + 0.0007 = 0.6671$.

This means that the probability of Z having a value < 0.432 is 0.6671.

How do you find $P(Z > 0.432)$? The table only gives the probabilities to the left of a value – that is, less than the value. However, the total area under the curve is 1, so you can calculate the difference.

So:

$P(Z > 0.432) = 1 - \Phi(0.432)$

$\qquad = 1 - 0.6671$

$\qquad = 0.3329$

Note that $P(Z < 0.432)$ $+ P(Z > 0.432) = 1$.

So the probability of Z having a value > 0.432 is 0.3329.

How do you find $P(Z < -0.432)$? The table does not give values of $Z < 0$.

You know that the normal distribution is symmetrical about the mean, so:

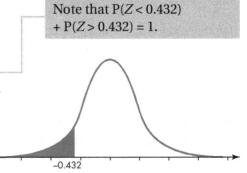

$P(Z < -0.432) = P(Z > 0.432)$

$\qquad = 1 - P(Z < 0.432)$

$\qquad = 1 - \Phi(0.432)$

$\qquad = 1 - 0.6671$

$\qquad = 0.3329$

So the probability of Z having a value < -0.432 is 0.3329.

Note that this is the same as $P(Z > 0.432)$.

Finally, how do you find $P(0.1 < Z < 0.31)$?

To find the probability (area) between 0.31 and 0.1, you need to find the probability that is less than 0.31 and subtract the probability that is less than 0.1.

$P(0.1 < Z < 0.31) = P(Z < 0.31) - P(Z < 0.1)$

$\qquad = \Phi(0.31) - \Phi(0.1)$

$\qquad = 0.6217 - 0.5398$

$\qquad = 0.0819$

> Good diagrams can be very useful when using the normal distribution tables to find probabilities.

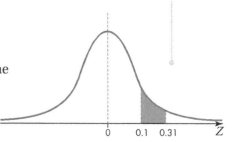

> It is important to show the stages in your workings. You must show what z-values you are looking up, as well as the values from the table.

What is the probability that Z will have a value greater than the mean but less than 1.5?

To find the probability (area) between the mean, 0, and 1.5, you need to find the probability that Z will have a value less than 1.5 and subtract the probability that Z will have a value less than 0.

$\qquad P(0 < Z < 1.5) = P(Z < 1.5) - P(Z < 0)$

The probability that Z is less than 0 is 0.5 since 0 is the mean and half of a normal distribution lies below the mean and half lies above it.

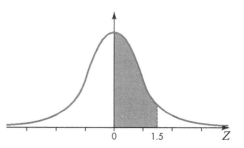

$P(0 < Z < 1.5) = P(Z < 1.5) - P(Z < 0)$

$\qquad = \Phi(1.5) - 0.5$

$\qquad = 0.9332 - 0.5$

$\qquad = 0.4332$

> The probability that Z is less than 0 always equals 0.5, as the curve is symmetrical at $Z = 0$.

Example 3

Find the following probabilities using tables:

a $P(Z \leqslant 0.347)$ **b** $P(Z < 0.205)$ **c** $P(Z > 0.6)$

Solution

a $P(Z \leqslant 0.347) = \Phi(0.347)$

z	0	1	2	3	4	5	6	7	8	9	1	2	3	4	5	6	7	8	9
																ADD			
0.0	0.5000	0.5040	0.5080	0.5120	0.5160	0.5199	0.5239	0.5279	0.5319	0.5359	4	8	12	16	20	24	28	32	36
0.1	0.5398	0.5438	0.5478	0.5517	0.5557	0.5596	0.5636	0.5675	0.5714	0.5753	4	8	12	16	20	24	28	32	36
0.2	0.5793	0.5832	0.5871	0.5910	0.5948	0.5987	0.6026	0.6064	0.6103	0.6141	4	8	12	15	19	23	27	31	35
0.3	0.6179	0.6217	0.6255	0.6293	0.6331	0.6368	0.6406	0.6443	0.6480	0.6517	4	7	11	15	19	22	26	30	34
0.4	0.6554	0.6591	0.6628	0.6664	0.6700	0.6736	0.6772	0.6808	0.6844	0.6879	4	7	11	14	18	22	25	29	32

$P(Z \leqslant 0.347) = 0.6331 + 0.0026$

$\qquad = 0.6357$

b $P(Z < 0.205) = \Phi(0.205)$

z	0	1	2	3	4	5	6	7	8	9	1	2	3	4	5	6	7	8	9
																ADD			
0.0	0.5000	0.5040	0.5080	0.5120	0.5160	0.5199	0.5239	0.5279	0.5319	0.5359	4	8	12	16	20	24	28	32	36
0.1	0.5398	0.5438	0.5478	0.5517	0.5557	0.5596	0.5636	0.5675	0.5714	0.5753	4	8	12	16	20	24	28	32	36
0.2	0.5793	0.5832	0.5871	0.5910	0.5948	0.5987	0.6026	0.6064	0.6103	0.6141	4	8	12	15	19	23	27	31	35
0.3	0.6179	0.6217	0.6255	0.6293	0.6331	0.6368	0.6406	0.6443	0.6480	0.6517	4	7	11	15	19	22	26	30	34
0.4	0.6554	0.6591	0.6628	0.6664	0.6700	0.6736	0.6772	0.6808	0.6844	0.6879	4	7	11	14	18	22	25	29	32

$P(Z < 0.205) = 0.5793 + 0.0019$

$\qquad = 0.5812$

c Draw a clear diagram and shade the required area.

$P(Z > 0.6) = 1 - P(Z < 0.6)$

$\qquad = 1 - \Phi(0.6)$

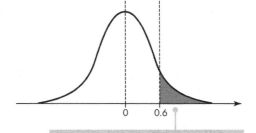

Z is greater than 0.6, so the shaded area lies to the right of 0.6.

z	0	1	2	3	4	5	6	7	8	9	1	2	3	4	5	6	7	8	9
																ADD			
0.0	0.5000	0.5040	0.5080	0.5120	0.5160	0.5199	0.5239	0.5279	0.5319	0.5359	4	8	12	16	20	24	28	32	36
0.1	0.5398	0.5438	0.5478	0.5517	0.5557	0.5596	0.5636	0.5675	0.5714	0.5753	4	8	12	16	20	24	28	32	36
0.2	0.5793	0.5832	0.5871	0.5910	0.5948	0.5987	0.6026	0.6064	0.6103	0.6141	4	8	12	15	19	23	27	31	35
0.3	0.6179	0.6217	0.6255	0.6293	0.6331	0.6368	0.6406	0.6443	0.6480	0.6517	4	7	11	15	19	22	26	30	34
0.4	0.6554	0.6591	0.6628	0.6664	0.6700	0.6736	0.6772	0.6808	0.6844	0.6879	4	7	11	14	18	22	25	29	32
0.5	0.6915	0.6950	0.6985	0.7019	0.7054	0.7088	0.7123	0.7157	0.7190	0.7224	3	7	10	14	17	20	24	27	31
0.6	0.7257	0.7291	0.7324	0.7357	0.7389	0.7422	0.7454	0.7486	0.7517	0.7549	3	7	10	13	16	19	23	26	29
0.7	0.7580	0.7611	0.7642	0.7673	0.7704	0.7734	0.7764	0.7794	0.7823	0.7852	3	6	9	12	15	18	21	24	27
0.8	0.7881	0.7910	0.7939	0.7967	0.7995	0.8023	0.8051	0.8078	0.8106	0.8133	3	5	8	11	14	16	19	22	25
0.9	0.8159	0.8186	0.8212	0.8238	0.8264	0.8289	0.8315	0.8340	0.8365	0.8389	3	5	8	10	13	15	18	20	23

$\qquad = 1 - 0.7257$

$\qquad = 0.2743$

Exercise 4.2B

For this exercise, use the correct notation and draw at least one diagram in each case.

1 Given that $Z \sim N(0, 1)$, draw a clear sketch diagram for:

   a $P(Z > a)$ where $a < 0$

   b $P(Z < a)$ where $a < 0$

   c $P(a < Z < b)$ where $a < 0$, $b > 0$

   d $P(a < Z < b)$ where $a, b < 0$, and $b > a$

   e $P(0 < Z < a)$ where $a > 0$.

2 A mathematics student was asked to write down the mathematical notation for the standard normal distribution.

   The student wrote $N \sim Z(1, 0^2)$.

   List everything that is wrong with what the student wrote and then write down the correct notation.

3 Find the probability that Z will be less than 1.423.

4 Find the probability that Z will be less than 0.87.

5 Find the probability that Z will be greater than 1.064.

6 Find the probability that Z will be less than –2.872.

7 Find the probability that Z will be greater than –1.326.

8 Find the probability that Z will be between 1.1 and 2.1.

9 Find the probability that Z will be between –1.325 and 1.218.

10 Find the probability that Z will be between –2.651 and –1.43.

11 Find $P(Z < -0.541)$.

12 Find $P(0.24 < Z < 1.102)$.

4.3 Non-standardised variables

You know that the area under a standard normally distributed curve (that is, with mean 0 and standard deviation 1) is equal to one. To find the probability that a random variable x is in any interval within the curve, you need to calculate the area of that interval. To find the area of any interval under any normal curve, convert to a z-score using the following formula:

$$z = \frac{\text{value} - \text{mean}}{\text{standard deviation}}$$

$$= \frac{x - \mu}{\sigma}$$

The horizontal scale of the curve of the standard normal distribution corresponds to z-scores.

If the random variables are converted into z-scores, the result will be the standard normal distribution. Following this conversion, the area that falls in the interval under the non-standard normal curve is the same as the area under the standard normal curve within the corresponding boundaries.

Example 4

Given the random variable $X \sim N(15, 4^2)$, find:

a $P(X < 25)$ **b** $P(X > 25)$ **c** $P(10 < X < 25)$

Solution

a Draw a clear diagram and shade the required area.

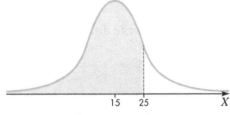

$$P(X < 25) = P\left(Z < \frac{25 - 15}{4}\right)$$

The variance is 4^2, so the standard deviation is 4.

$$= P(Z < 2.5)$$

$$= \Phi(2.5)$$

$$= 0.9938$$

So, the probability of X having a value < 25 is 0.9938.

b Draw a clear diagram and shade the required area.

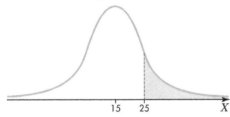

$$P(X > 25) = 1 - P(X < 25)$$
$$P(X > 25) = 1 - P\left(Z < \frac{25 - 15}{4}\right)$$

$$= 1 - P(Z < 2.5)$$

$$= 1 - \Phi(2.5)$$

$$= 1 - 0.9938$$

$$= 0.0062$$

So, the probability of X having a value > 25 is 0.0062.

You can also use the fact that probability sums to 1, along with the answer to part **a**, to work out $P(X > 25)$.

$$P(X > 25) = 1 - P(X < 25)$$
$$= 1 - 0.9938 = 0.0062$$

c Draw a clear diagram and shade the required area.

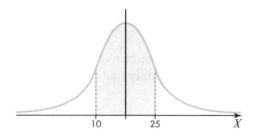

$$P(10 < X < 25) = P\left(\frac{10-15}{4} < Z < \frac{25-15}{4}\right)$$

$$= P(-1.25 < Z < 2.5)$$

$$= P(Z < 2.5) - P(Z < -1.25)$$

$$= \Phi(2.5) - \Phi(-1.25)$$

$$= \Phi(2.5) - (1 - \Phi(1.25))$$

$$= 0.9938 - (1 - 0.8944)$$

$$= 0.8882$$

So, the probability of X having a value between 10 and 25 is 0.8882.

Exercise 4.3A

For this exercise, present your working as shown in the examples above and use the correct notation.

1. Find the area of the indicated region under the standard normal curve.

a

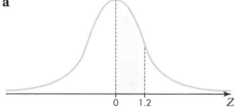

b

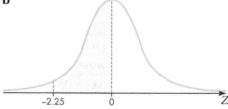

c

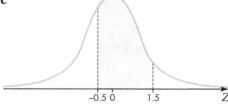

d

2 Given that $X \sim N(30, 100)$, find the following probabilities.

a $P(X < 35)$ **b** $P(X > 38.6)$ **c** $P(X > 20)$

d $P(35 < X < 40)$ **e** $P(15 < X < 32)$ **f** $P(17 < X < 19)$

3 The distribution of heights of 20-year-old men is modelled using the normal distribution with a mean of 177 cm and a standard deviation of 7 cm. Find the probability that the height of a randomly selected 20-year-old man is:

a under 172 cm **b** over 180 cm

c between 172 cm and 180 cm.

4 The lengths of a population of caterpillars are approximately normally distributed with a mean 10.8 cm and a standard deviation 1.6 cm. Find the probability that the length of a randomly selected caterpillar is:

a more than 15.5 cm

b less than or equal to 10 cm

c between 9.2 and 11.2 cm.

5 When Tim makes hot chocolate to drink, he spoons instant powder onto a spoon and places it into a mug. The weight of powder in grams may be modelled by the normal distribution with a mean of 6 g and a standard deviation of 1 g. Tim has found that if he uses more than 8 g Jenni says it's too strong and if he uses less than 4.5 g she says it's too weak. Find the probability that he makes the hot chocolate:

a too weak **b** too strong **c** an acceptable strength.

6 A particular breed of hens produces eggs with a mean mass of 60 g with a standard deviation of 4 g. Mass is found to be normally distributed.

The eggs are classified as small, medium, large or extra large depending on their mass, as follows:

Classification	Mass
small	less than 55 g
medium	between 55 g and 65 g
large	between 65 g and 70 g
extra large	greater than 70 g

a Find the proportion of eggs that are classified as small.

b Find the proportion of eggs that are classified as medium.

c Find the proportion of eggs that are classified as large or extra large.

d If the classification for medium eggs is changed to between 55 g and 68 g, by how much does the proportion of eggs classified as medium increase?

7 The heights of college students are normally distributed with a mean of 168 cm and a standard deviation of 4.6 cm. A student is selected at random.

a Find the probability that the student is shorter than 158 cm.

George is looking for a student whose height is between 165 cm and 170 cm for his art project.

b Find the proportion of students that meets George's criteria.

8 The police regularly monitor the speeds of cars on a section of motorway and it is found that the speeds are normally distributed. The mean speed is 68.5 km/h with a standard deviation of 5 km/h.

a Find the proportion of motorists who break the speed limit (70 km/h).

b A motorist believes that people aren't fined unless they are 10% over the speed limit.

What proportion of motorists will be fined on this basis?

c A motorist is given a fixed penalty fine for driving at 85 km/h.

What percentage of motorists exceed this speed?

> You can find some commonly used z values using the critical values for the normal distribution table.

9 The volume of skimmed milk in a 2-litre bottle may be modelled by a normal distribution with a mean of 2.025 litres and standard deviation 0.015. Find the probability that the volume of a random selected 2-litre skimmed milk bottle is:

a less than 2.05 litres

b more than 2 litres.

10 Bags of sweets are produced with a mean weight 250 g. Quality control checks show that 0.5% of bags are rejected because their weight is less than 235 g. It can be assumed that the weight is normally distributed.

a Find the standard deviation of the weights of bags of sweets.

b Hence find the proportion of bags that weigh more than 260 g.

Mathematics in life and work: Group discussion

You have set up a new factory machine to make needles. You know that the length of needles has a mean of 3.5 cm and a standard deviation of 0.2 cm.

1 What percentage of needles made by the machine will be less than 3.8 cm?

2 The acceptable length for a needle is between 2.95 and 3.8 cm. If the machine makes 300 needles, how many would you expect to be rejected?

3 How could you change the setting of the mean of the machine to maximise the number of acceptable needles?

4 A different machine has a mean of 3.4 cm and a standard deviation of 0.1 cm. It is known that as you change the mean by 0.1, the standard deviation increases by 0.05. Which machine would maximise the number of acceptable needles?

Finding values of variables from known probabilities

Sometimes you may know the probability of an event and then need to work backwards to find the corresponding value X.

Example 5

A random variable X is normally distributed with a mean 9 and standard deviation 0.27.

a Find the value of x so that $P(X < x) = 0.7541$.

b Find the value of a so that $P(X < a) = 0.25$.

c Find the value of b so that $P(-b < \text{mean} < b) = 0.6$.

Solution

a First, write down the distribution in the question as a standardised normal distribution.

$P(Z < z) = 0.7541$ where $z = \dfrac{x - 9}{0.27}$

$z = \Phi^{-1}(0.7541)$

Recall that $\dfrac{x - \mu}{\sigma} = z.$

To find the z-value from a given probability you need to use the tables backwards. First find the closest probability that is lower than 0.7541:

z	0	1	2	3	4	5	6	7	8	9	1	2	3	4	5	6	7	8	9
															ADD				
0.0	0.5000	0.5040	0.5080	0.5120	0.5160	0.5199	0.5239	0.5279	0.5319	0.5359	4	8	12	16	20	24	28	32	36
0.1	0.5398	0.5438	0.5478	0.5517	0.5557	0.5596	0.5636	0.5675	0.5714	0.5753	4	8	12	16	20	24	28	32	36
0.2	0.5793	0.5832	0.5871	0.5910	0.5948	0.5987	0.6026	0.6064	0.6103	0.6141	4	8	12	15	19	23	27	31	35
0.3	0.6179	0.6217	0.6255	0.6293	0.6331	0.6368	0.6406	0.6443	0.6480	0.6517	4	7	11	15	19	22	26	30	34
0.4	0.6554	0.6591	0.6628	0.6664	0.6700	0.6736	0.6772	0.6808	0.6844	0.6879	4	7	11	14	18	22	25	29	32
0.5	0.6915	0.6950	0.6985	0.7019	0.7054	0.7088	0.7123	0.7157	0.7190	0.7224	3	7	10	14	17	20	24	27	31
0.6	0.7257	0.7291	0.7324	0.7357	0.7389	0.7422	0.7454	0.7486	0.7517	0.7549	3	7	10	13	16	19	23	26	29
0.7	0.7580	0.7611	0.7642	0.7673	0.7704	0.7734	0.7764	0.7794	0.7823	0.7852	3	6	9	12	15	18	21	24	27
0.8	0.7881	0.7910	0.7939	0.7967	0.7995	0.8023	0.8051	0.8078	0.8106	0.8133	3	5	8	11	14	16	19	22	25
0.9	0.8159	0.8186	0.8212	0.8238	0.8264	0.8289	0.8315	0.8340	0.8365	0.8389	3	5	8	10	13	15	18	20	23

The closest value is 0.7517. Now look in the thousandths table to find the value that needs to be added on to 0.7517 to get as close as possible to 0.7541. This gives a z-value of 0.687.

$\dfrac{x - 9}{0.27} = 0.687$

$x - 9 = 0.687 \times 0.27$

$x = 0.18549 + 9$

$x = 9.185$

Note the z-value is rounded to three decimal places, it is not the exact value.

Probability of $X < 9.185$ is 0.7541, so $X = 9.185$.

b $P(X < a) = 0.25$ written as a standardised normal distribution is $P(Z < z) = 0.25$ where

$z = \dfrac{a - 9}{0.27}$. The table only gives probabilities with 0.5 or greater.

You need to draw a clear diagram and label the area that equals 0.25.

$\Phi^{-1}(-z) = 1 - 0.25$

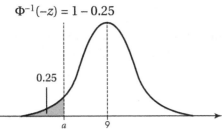

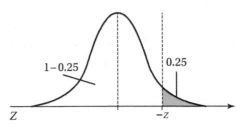

z	0	1	2	3	4	5	6	7	8	9	1	2	3	4	5	6	7	8	9
														ADD					
0.0	0.5000	0.5040	0.5080	0.5120	0.5160	0.5199	0.5239	0.5279	0.5319	0.5359	4	8	12	16	20	24	28	32	36
0.1	0.5398	0.5438	0.5478	0.5517	0.5557	0.5596	0.5636	0.5675	0.5714	0.5753	4	8	12	16	20	24	28	32	36
0.2	0.5793	0.5832	0.5871	0.5910	0.5948	0.5987	0.6026	0.6064	0.6103	0.6141	4	8	12	15	19	23	27	31	35
0.3	0.6179	0.6217	0.6255	0.6293	0.6331	0.6368	0.6406	0.6443	0.6480	0.6517	4	7	11	15	19	22	26	30	34
0.4	0.6554	0.6591	0.6628	0.6664	0.6700	0.6736	0.6772	0.6808	0.6844	0.6879	4	7	11	14	18	22	25	29	32
0.5	0.6915	0.6950	0.6985	0.7019	0.7054	0.7088	0.7123	0.7157	0.7190	0.7224	3	7	10	14	17	20	24	27	31
0.6	0.7257	0.7291	0.7324	0.7357	0.7389	0.7422	0.7454	0.7486	0.7517	0.7549	3	7	10	13	16	19	23	26	29
0.7	0.7580	0.7611	0.7642	0.7673	0.7704	0.7734	0.7764	0.7794	0.7823	0.7852	3	6	9	12	15	18	21	24	27
0.8	0.7881	0.7910	0.7939	0.7967	0.7995	0.8023	0.8051	0.8078	0.8106	0.8133	3	5	8	11	14	16	19	22	25
0.9	0.8159	0.8186	0.8212	0.8238	0.8264	0.8289	0.8315	0.8340	0.8365	0.8389	3	5	8	10	13	15	18	20	23

This is not the exact value. However, from the table, the closest value to 0.75 is 0.7499 (= 0.7486 + 0.0013).

$\Phi^{-1}(-z) = 0.75$

> By symmetry, $P(Z > -z) = 0.25$. Therefore $P(Z < z) = 1 - 0.25$.

$-z = 0.674$

$z = -0.674$

$\dfrac{a - 9}{0.27} = -0.674$

> Alternatively, use the critical values for the normal distribution table. The first column shows that $z = 0.674$ when probability is 0.75.

$a = -0.674 \times 0.27 + 9$

$= 8.818$

Probability of $X < 8.818$ is 0.25 so $a = 8.818$.

c $P(-b < X < b) = 0.6$ can be written as $P(-z < Z < z) = 0.6$.

$P(0 < Z < z) = 0.3$

So:

$P(Z < z) = P(Z < 0) + P(0 < Z < z)$ where $z = \dfrac{b - 9}{0.27}$

$= 0.5 + 0.3$

$= 0.8$

$\Phi^{-1}(0.8) = z$

$z = 0.842$

From the graph, you know that:

$\dfrac{(b - 9)}{0.27} = 0.842$

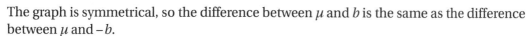

$b - 9 = 0.842 \times 0.27$

$b = 9.227$

The graph is symmetrical, so the difference between μ and b is the same as the difference between μ and $-b$.

Therefore:

$-b = 9 - (9.227 - 9)$

$-b = 8.773$

Probability of $8.773 < X < 9.227$ with a mean of 9 and standard deviation of 0.27 is 0.6.

Exercise 4.3B

1. Find the z-value that corresponds to the following:

   a $P(Z < z) = 0.7642$ b $P(Z < z) = 0.638$ c $P(Z > z) = 0.195$

   d $P(Z > z) = 0.375$ e $P(Z > z) = 0.8531$ f $P(Z > z) = 0.6$

   g $P(Z < z) = 0.372$ h $P(0 < Z < z) = 0.45$ i $P(-z < Z < z) = 0.8$

2. A factory produces bags of sugar. The actual weights may be modelled by a normal distribution with mean 498.7 g and standard deviation 7.3 g. Bags are rejected if their weights lie in the bottom 15% or the top 10% of the distribution. What weights are these?

3. Assume that the results of a test are normally distributed with a mean of 19 and a standard deviation of 2.4. Let X be the distribution of test scores. Find a and b to 2 decimal places, where:

   a $P(X > a) = 0.432$ b $P(X < b) = 0.205$.

4. Given that $X \sim N(30, 100)$, find x where:

   a $P(X < x) = 0.99$ b $P(X < x) = 0.9798$ c $P(X > x) = 0.1949$

   d $P(X < x) = 0.75$ e $P(X < x) = 0.35$ f $P(X > x) = 0.05$.

5. A manufacturer produces a new fluorescent light bulb and claims that it has an average lifetime of 4000 hours with a standard deviation of 375 hours. Assume that the lifetime of the new fluorescent light bulbs is normally distributed.

   > The critical values for the normal distribution table can be used to find the value of z.

   Find the lifetime such that 90% of light bulbs will last for less than this duration.

6. A double bed is the right length for 92.9% of men.

   The heights of men in the UK are normally distributed with mean height 1.753 m and standard deviation 0.1 m.

   What is the greatest height a man can be to fit onto a double bed?

7. The police regularly monitor the speeds of cars on a section of motorway and it is found that the speeds are normally distributed. The mean speed is 68.5 km/h^{-1} with a standard deviation of 5 km/h^{-1}.

   Find the symmetrical range about the mean within which 80% of the speeds lie.

8. An examination has a mean mark of 80 and a standard deviation of 12. The marks can be assumed to be normally distributed.

   a Explain why a normal distribution would be an approximate distribution.

   b What is the lowest mark needed to be in the top 25% of students taking this examination?

   c Between which two marks will the middle 95% of students lie?

   d 200 students take this examination. Calculate the number of students likely to score 90 or more.

9 The weight, X grams, of the contents of a tin of soup can be modelled by a normal random variable with a mean of 190 g and a standard deviation of 2.5.

   a What is the lowest weight , to 1 decimal place, of a tin of soup in the heaviest 20% of soup tins?

   b What is the range in weights, to 1 decimal place, of the middle 40% of the soup tins around the mean weight?

   c 100 tins were selected at random. Calculate the number of tins likely to be 195 g or more.

10 During 2018, the volume, X litres, of unleaded petrol purchased per visit at a petrol station by private car customers could be modelled by a normal distribution with a mean of 35 and a standard deviation of 8.

   a Find the probability that a private car customer purchases less than 20 litres of petrol.

   b Given that during May 2018 unleaded petrol cost $1.55 per litre, calculate the probability that the unleaded petrol bill for a visit during May 2018 by a private car customer exceeded $45.

   c Give a reason, in context, why the model $N(35, 8^2)$ is unlikely to be valid for a visit by any customer purchasing fuel at this petrol station during May 2018.

Determining the mean or standard deviation

There are instances where a random variable is known to be normally distributed but the value of the mean or the standard deviation (or both) is not known. When you have information about the probabilities, it is possible to determine the unknown parameters.

Example 6

A farm that produces potatoes models the weights (in g) of a particular brand of potato by a random variable, $J \sim N(\mu, 25)$. The processing plant observes that 2% of the potatoes weigh more than 350.7 g. Determine the mean, μ.

Solution

Draw a sketch of the given information:

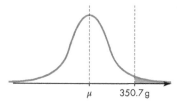

You need to find the area of the shaded region. The question says 2% of the potatoes have a weight of more than 350.7 g, therefore the area is 2% to the right.

> The total area under the graph equals 1.

Therefore, $P(Z > z) = 0.02$ with $z = \dfrac{350.7 - \mu}{5}$.

$P(Z < z) = 0.98$

$\Phi^{-1}(0.98) = 2.054$

This gives $P(Z > 2.054) = 0.02$.

The standardised value of 350.7 satisfies the equation:

$$\frac{350.7 - \mu}{5} = 2.054$$

Multiply by 5 to give $350.7 - \mu = 10.27$.

Solving this for $\mu = 340.43 = 340.4\,$g (to 1 d.p.)

Example 7

A new process is introduced at the farm. This changes the variance, so the weight of the potatoes is now modelled by the random variable $M \sim N(340, \sigma^2)$. The processing plant reveals that 90% of them have a weight above 322.3 g. Determine the variance of the random variable.

Solution

Draw a sketch:

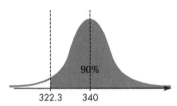

You now need to find the area of the shaded region. The question says that 90% of the potatoes have a weight of more than 322.3 g, therefore the area is 90% to the right.

Therefore, $P(Z > z) = 0.9$ with $z = \dfrac{322.3 - 340}{\sigma}$.

By symmetry, $P(Z < -z) = 0.9$.

> Alternatively, the critical values for the normal distribution table can be used to find z values.

$-z = \Phi^{-1}(0.9) = 1.282$

$z = -1.282$

This gives $P(Z > -1.282) = 0.90$.

The standardised value of 322.3 satisfies the equation:

$$\frac{322.3 - 340}{\sigma} = -1.282$$

Multiply by σ to give $322.3 - 340 = -1.282\sigma$.

$-17.7 = -1.282\sigma$

Solving this for $\sigma = 13.81$.

So the variance, $\sigma^2 = 190.62\,$g.

Example 8

The weight of potatoes put into bags by the processing plant is modelled by a normal distribution with a mean of μ g and a standard deviation of σ g.

The processing plant observe that in the bags, 83% contain less than 354.8 g and 8.5% contain less than 343.1 g. Determine μ and σ.

Solution

As there are two pieces of information, two sketches are required.

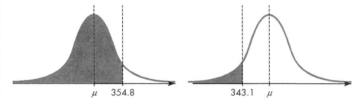

You need to find the area of the shaded regions.

Area 1: The question says that 83% of the potatoes have a weight of less than 354.8 g, therefore the area is 83%. This can be written as $P(Z < z) = 0.83$ where:

$$z = \frac{354.8 - \mu}{\sigma}$$

and $\Phi^{-1}(0.83) = 0.954$

This gives: $P(Z < 0.954) = 0.83$

The standardised value of 354.8 satisfies the equation:

$$\frac{354.8 - \mu}{\sigma} = 0.954$$

Area 2: The question says that 8.5% of the potatoes have a weight of less than 343.1 g, therefore the area is 8.5%. This can be written as $P(Z < z) = 0.085$ where $z = \frac{343.1 - \mu}{\sigma}$ By symmetry, we know that $P(Z > -z) = 0.085$ and so $P(Z < -z) = 1 - 0.085$.

$$-z = \Phi^{-1}(0.915) = 1.372$$

$$z = -1.372$$

This gives $P(Z < -1.372) = 0.085$.

The standardised value of 343.1 satisfies the equation:

$$\frac{343.1 - \mu}{\sigma} = -1.372$$

You can now rearrange both of these equations to set up a pair of simultaneous equations in μ and σ.

$$354.8 - \mu = 0.954\sigma$$

$343.1 - \mu = -1.372\sigma$

Subtract both equations to eliminate μ.

$11.7 = 2.326\sigma$

$\sigma = 5.03\,\mathrm{g}$

Therefore, substituting back into one of the two derived equations:

$354.8 - \mu = 0.954\sigma$

$= 4.799$

$\mu = 350\,\mathrm{g}$

Exercise 4.3C

(PS) **1** Assume that the results of a test are normally distributed with a mean of 43 and a standard deviation of σ. Let X be the distribution of test scores.

Find σ if the probability of getting a score above 48 is 0.2.

(PS) **2** The length of the frame of a bike may be modelled by a normal distribution with standard deviation 13 mm. 11% of the components are longer than 47 cm.

Calculate the mean length of a frame.

(PS) **3** The volume of the contents of a can of fizzy drink may be modelled by a normal distribution. 18% have a volume of more than 332.91 ml and 72% have a volume of more than 325.42 ml. Find the mean and standard deviation of the volume.

(PS) **4** The quartiles of a normal distribution are known to be 9.92 and 12.24. Find the mean and standard deviation of the distribution to 4 significant figures.

5 Decorative pebbles are sold individually with a mean diameter of 15.5 cm. A sales manager finds that 1% of the pebbles are returned to the manufacturer because their diameter is greater than 18 cm.

a Find the standard deviation of the diameter of a pebble.

b Hence find the proportion of pebbles that are smaller than 14 cm.

6 A bakery sells square fruit cakes, the masses of which can be modelled by a normal distribution with a mean of 500 grams and a standard deviation of σ.

The manager requests that at most, 10 per cent of these cakes should weigh less than 480 grams. Find:

a the maximum value of σ

b the probability that the cake is more than 508 grams.

PS **7** Engineers make certain components for aeroplanes. The company believes that the time taken to make a particular type of component may be modelled by a normal distribution with a mean of 90 minutes and a standard deviation of three minutes.

Assuming the company's beliefs to be true, find the probability that the time taken to make one of these components, selected at random, was:

a over 95 minutes

b under 87 minutes

c between 87 and 95 minutes.

Stan believes that the company is allowing too long for the job and times himself manufacturing the component. He finds that only 10% of the components take him over 85 minutes, and that 20% take him less than 65 minutes.

d Estimate Stan's mean and standard deviation.

8 A power lifter knows from experience that he can deadlift at least 140 kg once in every five attempts. He also knows that he can deadlift at least 130 kg on eight out of 10 attempts. Find the mean and standard deviation of the weights the powerlifter can dead lift.

9 The time taken for a chemical reaction to occur can be modelled by a normal distribution with a mean of 12 minutes and a standard deviation of two minutes. Find the probability that the time taken for a randomly selected reaction was:

a under 11 minutes

b over 13.5 minutes

c under 10 or over 13 minutes.

A student believes that the higher temperature reduces the reaction time. She carries out experiments in a water bath, which has constant temperature of 36 °C. She finds that only 5% of the experiments take over 13 minutes and 12% take less than 10 minutes.

d Estimate the mean and standard deviation of the reaction time at 36 °C.

10 Mr Wang attempts a Sudoku puzzle every day. The time taken, X minutes, to complete the puzzle may be modelled by a normal distribution with mean 16 and standard deviation 4.5.

a Calculate the probability that he takes more than 18 minutes or less than 12 minutes to complete the Sudoku.

b What length of time would be enough for Mr Wang to finish the Sudoku on 90% of days?

Mr Wang bought another Sudoku puzzle book and finds that on 99% of occasions he completes the puzzle within 19 minutes.

c Assuming that the time taken, Y minutes, to complete the Sudoku puzzle has the distribution N(16, σ^2), find the value of σ.

Mathematics in life and work: Group discussion

You are an engineer for a factory that manufactures rivets (metal pins to join metal sheets together). The machine is set to a mean diameter of 3 mm. It is noticed that 90% of rivets are between 2.75 mm and 3.25 mm.

1 What is the standard deviation of the diameter of the rivets?

2 The rivets fit into holes that have mean diameter of 3.1 mm and standard deviation of 0.6 mm. How could you change the settings of your machine to maximise the number of rivets that fit?

As the machine gets older you think that the diameter is changing. You think the mean and the standard deviation may have changed due to wear and tear of the machine.

3 How could you check to see if this is true?

4.4 Using the normal distribution to approximate the binomial distribution

In **Chapter 3 Discrete random variables** you learned about the binomial distribution. You are now going to use **binomials** with the normal distribution.

A fair coin has the same probability of landing on heads as it does on tails. If you want to know the probability of getting exactly nine heads out of 15 flips you could use the binomial distribution.

> The mean is $\mu = np = 15 \times 0.5 = 7.5$.

> The variance is $\sigma^2 = np(1 - p) = 15 \times 0.5 \times 0.5 = 3.75$.

> The standard deviation is $\sigma = \sqrt{3.75} = 1.93645$.

To find the exact probability of flipping nine heads out of 15 you would use $X \sim B(n, p)$ on your calculator or $^{15}C_9 \times 0.5^9 \times 0.5^6 = 0.15274$.

So P(exactly nine heads) = 0.15274.

The z-score for exactly nine heads is $\frac{(9 - 7.5)}{1.9365} = 0.775$ standard deviations above the mean of the distribution.

$$z = \frac{\text{value} - \text{mean}}{\text{standard deviation}}$$
$$= \frac{x - \mu}{\sigma}$$

You already know that the probability of any one specific point is 0, as the normal distribution is a continuous distribution. However, the binomial distribution is a discrete probability distribution, so you need to use something called a **continuity correction**.

To find P(exactly nine heads) you need to consider any value from 8.5 to 9.5. Once you are considering a range, you will have an interval, so you can work out the area under a normal curve from 8.5 to 9.5.

$$P(8.5 < X < 9.5) = P\left(\frac{8.5 - 7.5}{1.9365} < Z < \frac{9.5 - 7.5}{1.9365}\right)$$

$$= P(0.516 < Z < 1.033)$$

$$= \Phi(1.033) - \Phi(0.516)$$

$$= 0.8492 - 0.6970 = 0.1522$$

The area of the interval is 0.152, which is the approximation of the binomial probability. For these parameters, the approximation is very accurate.

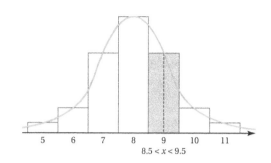

$8.5 < x < 9.5$

You may use the normal distribution as an approximation for the binomial $B(n, p)$ (where n is the number of trials, each having probability p of success) when n is sufficiently large to allow $np \geqslant 5$ and $nq \geqslant 5$.

> Remember from **Chapter 3**
> **Discrete random variables**
> that $q = 1 - p$.

This ensures that the distribution is reasonably symmetrical and not skewed at either end.

The parameters for the normal distribution are then:

> mean $= \mu = np$

> variance $= \sigma^2 = np(1 - p)$

> standard deviation $= \sqrt{np(1 - p)}$

KEY INFORMATION

If $np \geqslant 5$ and $n(1 - p) \geqslant 5$ then the binomial random variable X can be approximated by $Y \sim N(np, np(1 - p))$.

Calculating the probability of a continuous random variable for a specific value of x will be zero, because a range is required to have a non-zero area under the curve. Since you are using a continuous random variable to approximate a discrete random variable, a continuity correction can be used to introduce a range. Typically, this is achieved by considering the range $(x \pm 0.5)$.

For example, if you wanted the probability of x, then in the continuous approximation you would need to find $P(x - 0.5 < x < x + 0.5)$.

In deciding whether to round up or down when applying a continuity correction, the following list is a useful guide:

> If $P(X = n)$ use $P(n - 0.5 < X < n + 0.5)$

> If $P(X > n)$ use $P(X > n + 0.5)$

> If $P(X \leqslant n)$ use $P(X < n + 0.5)$

> If $P(X < n)$ use $P(X < n - 0.5)$

> If $P(X \geqslant n)$ use $P(X > n - 0.5)$

Example 9

In two surveys, respondents answer 'yes' or 'no'. In each case, decide whether you can use the normal distribution to approximate X, the number of people who answer 'yes'. If you can, find the mean and standard deviation. If you cannot, explain why.

a 34% of people in Lincoln say that they are likely to make a New Year's resolution. You randomly select 15 people in Lincoln and ask each if they are likely to make a New Year's resolution.

b 6% of people in Lincoln who made a New Year's resolution resolved to exercise more. You randomly select 65 people in Lincoln who made a resolution and ask each if they have resolved to exercise more.

Solution

a In this binomial experiment, $n = 15$, $p = 0.34$, and $(1 - p) = 0.66$. So, $np = (15)(0.34) = 5.1$ and $nq = (15)(0.66) = 9.9$.

Because np and $n(1 - p)$ are greater than 5, you can use the normal distribution with $\mu = 5.1$ and $\sigma = \sqrt{np\,(1 - p)} = 1.83$.

b In this binomial experiment, $n = 65$, $p = 0.06$, and $(1 - p) = 0.94$. So, $np = (65)(0.06) = 3.9$ and $nq = (65)(0.94) = 61.1$. You should notice here that $np < 5$. This means that you cannot use the normal distribution to approximate the distribution of X.

Example 10

Use a continuity correction to convert each of the following binomial intervals to a normal distribution interval:

a the probability of getting between 170 and 290 successes inclusive

b the probability of at least 58 successes

c the probability of getting less than 23 successes.

Solution

a The discrete mid-point values are 170, 171, ... 290, so in the continuous normal distribution the probability is $169.5 < x < 290.5$.

b The discrete mid-point values are 58, 59, 60, ... Therefore, in the continuous normal distribution the probability is $x > 57.5$.

c The discrete mid-point values are ... 20, 21, 22, so in the continuous normal distribution the probability is $x < 22.5$.

Example 11

You are told that 38% of people in the UK admit that they look at other people's mobile phones. You randomly select 200 people in the UK and ask each if they look at other people's mobile phones. If the statement is true, what is the probability that at least 70 will say yes?

Solution

Because $np = 200 \times 0.38 = 76$ and $n(1 - p) = 200 \times 0.62 = 124$, the binomial variable X is approximately normally distributed with:

> $\mu = np = 76$

> $\sigma = \sqrt{(200 \times 0.38 \times 0.62)} = 6.86$

Using a continuity correction, you can rewrite the discrete probability $P(X \geqslant 70)$ as the continuous probability function $P(X > 69.5)$. The graph shows a normal curve with $\mu = 76$ and $\sigma = 6.86$ and a shaded area to the right of 69.5.

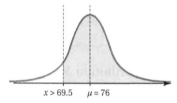

$x > 69.5 \qquad \mu = 76$

The z-score that corresponds to 69.5 is $z = \dfrac{(69.5 - 76)}{6.86} = -0.948$.

So, the probability that at least 70 will say yes is:

$P(X > 69.5) = P(Z > -0.948)$

$\qquad\qquad = P(Z < 0.948)$

$\qquad\qquad = 0.8284$

Example 12

A survey reports that 95% of teenagers eat a burger each week. You randomly select 200 teenagers and ask each whether they have eaten a burger this week. Using a normal distribution, what is the probability that exactly 194 will say yes?

Solution

Because $np = 200 \times 0.95 = 190$ and $n(1 - p) = 200 \times 0.05 = 10$, the binomial variable x is approximately normally distributed with:

> $\mu = np = 190$

> $\sigma = \sqrt{(200 \times 0.95 \times 0.05)} = 3.0822$

Using a continuity correction, you can rewrite the discrete probability P(X = 194) as the continuous probability function P(193.5 < X < 194.5).

The first z-score that corresponds to 193.5 and 194.5 would be $z = \dfrac{193.5 - 190}{3.0822} = 1.136$.

The second z-score would be $z = \dfrac{194.5 - 190}{3.0822} = 1.460$.

So, the probability that exactly 194 teenagers will say they have eaten a burger is:

$$\text{P}(193.5 < X < 194.5) = \text{P}(1.136 < Z < 1.460)$$
$$= \text{P}(Z < 1.460) - \text{P}(Z < 1.136)$$
$$= 0.9279 - 0.8720$$
$$= 0.0559$$

There is a probability of about 0.06 that exactly 194 of 200 teenagers will say they have eaten a burger this week.

Exercise 4.4A

1 You decide to use the normal distribution to approximate the binomial distribution. You want to know the probability of getting exactly 16 tails out of 20 coin flips.

    a Calculate the mean number of tails in 20 coin flips.

    b Calculate the variance and standard deviation.

    c Calculate the probability using the normal distribution of 16 tails out of 20 coin flips.

2 108 people took a test for which the probability of passing is 0.88. Let X be the number of people who passed the test. Using the normal distribution to approximate the binomial distribution, determine the probabilities:

    a P(X < 100)   **b** P(85 < X < 105).

3 The discrete random variable X has a binomial distribution with n = 200 and p = 0.32. Determine, by using a suitable approximation, the probabilities that:

    a P(X < 130)   **b** P(X > 50)

    c P(X > 75)   **d** P(75 < X < 130).

4 **a** State the conditions under which a binomial probability model can be well approximated by a normal model.

    T is a random variable with the distribution B(15, 0.4).

    b Rahman uses the binomial distribution to calculate the probability that T < 5 and gives his answer to 4 significant figures. What answer does he get?

    c Cynthia uses a normal distribution to calculate an approximation for the probability that T < 5 and gives her answer to 4 significant figures. What answer does she get?

d Assuming that Cynthia has worked out everything correctly, calculate the percentage error in her calculation.

5 A restaurant serves lunch every day. The manager claims that half of the customers have desserts. 120 customers were randomly selected one day, using the normal approximation to the binomial, what is the probability that at least 55 customers had desserts?

6 In a board game, the probability of a counter landing in a lucky square is 0.35. Calculate the probability that in 50 games the counter lands in the lucky square exactly 20 times:

a using the binomial distribution

b using the normal approximation to the binomial.

7 A local shopkeeper knows, from past records, that 20% of customers buy a bag of sweets from those displayed next to the till. 200 customers were randomly selected, using the normal approximation to the binomial. Find the probability that:

a at least 29 customers bought a bag of sweets

b exactly 45 customers bought a bag of sweets.

8 A bag contains a large number of sweets of which 12% are strawberry flavour. 30 sweets are randomly picked and the number with a strawberry flavour is recorded.

a Find the probability that less than three strawberry-flavoured sweets were selected.

Another 30 sweets are randomly picked and the number with a strawberry flavour is recorded.

b Using a normal distribution approximation, estimate the probability that the total number of strawberry-flavoured sweets in the combined selection of size 60 is less than 9.

9 A popular brand of moisturiser claims that 72% of people using it see noticeable effects after just two weeks.

Assuming the manufacturer's claim is correct for the population using the moisturiser, calculate the probability that at least 13 of a random sample of 15 people using the moisturiser can see noticeable effects:

a using the binomial distribution

b using the normal approximation to the binomial.

c Comment on the agreement or disagreement between your two values.

d Would the agreement be better or worse if the proportion had been 85% instead of 72%?

10 A multiple-choice theory test consists of 35 questions. For each question, the candidate is required to tick one of five possible answers. Exactly one answer to each question is correct. A correct answer gains one mark and an incorrect answer gains no marks.

One candidate guesses every answer without looking at the question first.

a What is the probability that the candidate gets a particular answer correct?

b Calculate the mean and variance of the number of questions answered correctly.

c The panel of examiners want to ensure that no more than 1% of candidates who guess in this way pass the examination. Use the normal approximation to the binomial, working to 3 decimal places, to establish the minimum pass mark that meets this requirement.

SUMMARY OF KEY POINTS

› A random variable X that is normally distributed, with mean μ and variance σ^2, is notated $X \sim N(\mu, \sigma^2)$.

› The standard normal distribution is defined as $Z \sim N(0, 1^2)$, with a mean of 0 and a standard deviation of 1.

› To standardise a variable x to become standard normally distributed, use $z = \dfrac{x - \mu}{\sigma}$.

› The normal distribution can be used to approximate discrete distributions but continuity corrections are required.

› The binomial distribution $B(n, p)$ can be approximated by $N(np, np(1 - p))$ provided that $np \geqslant 5$ and $nq \geqslant 5$.

EXAM-STYLE QUESTIONS

1   The mass of sugar in a 1 kg bag may be assumed to have a normal distribution with mean 1005 g and standard deviation 2 g. Find the probability that:

   a a 1 kg bag will contain less than 1000 g of sugar

   b a 1 kg bag will contain more than 1007 g of sugar.

   Two 1 kg bags are chosen at random.

   c Find the probability that they both contain between 1000 g and 1007 g of sugar.

2   The heights of the UK adult female population are normally distributed with mean 164.5 cm and standard deviation 8.75 cm.

   a Find the probability that a randomly chosen adult female is taller than 160 cm.

   Nancy is a student in Year 7. She is at the 45th percentile for her height.

   b Assuming that Nancy remains at the 45th percentile, estimate her height as an adult.

3   The mass of people using a lift is normally distributed with mean 72 kg and standard deviation 10 kg. The lift has a maximum allowed load of 320 kg. If four people from this population are in the lift, determine the probability that the maximum load is exceeded. State any assumption you made during the calculation.

4   A town gets, on average, 82 mm of rain per month with a standard deviation of 6 mm. This can be modelled by normal distribution.

   a Find the expected number of months in a year that get more than 90 mm of rain.

   A different town has μ mm of rain per month with a standard deviation of 5 mm. It is known that the probability of getting less than 53 mm of rain in a month is 0.38.

   b Find the value of μ.

   c Find the probability that exactly four months in a year get less than 60 mm of rain in the second town.

5 A university laboratory is lit by a large number of light bulbs with lifetimes modelled by a normal distribution with a mean of 1000 hours and standard deviation of 110 hours. The bulbs remain switched on continuously.

    a What proportion of light bulbs have lifetimes that exceed 900 hours?

    b What proportion of light bulbs have lifetimes that exceed 1200 hours?

    c Find the probability that a bulb will last more than 900 hours but less than 1000 hours.

    The university replace the light bulbs periodically after a fixed interval.

    d To the nearest day, how long should this interval be if, on average, 1% of the light bulbs are to burn out between successive replacement times?

 6 A shopkeeper finds that 20% of customers buy an item that is displayed next to the checkout. In a randomly chosen day, 30 customers visit his shop.

    a Describe fully the distribution of X, the number of customers who buy an item displayed next to the checkout, and a suitable approximation. Justify your answer.

    b Using the approximation, find the probability that at least four people buy an item displayed next to the checkout.

    On the following two days, 30 customers visit the shop each day.

    c Find the probability that at least four people buy an item displayed next to the checkout on each of these two days.

 7 A factory produces blades for wind turbines. The lengths of these blades are normally distributed with 33% of them measuring 212.6 cm or more and 12% of them measuring 211.8 cm or less.

    a Write down simultaneous equations for the mean and standard deviation of the distribution and solve to find the values. Hence estimate the proportion of blades that measure 212 cm or more.

    b The blades are acceptable if they measure between 211.8 cm and 212.8 cm. What percentage is rejected as being outside of the acceptable range?

 8 A machine is used to fill oil into gearboxes with a nominal volume of 3.54 litres. Suppose the machine delivers a quantity of oil that is normally distributed with a mean of 3.58 litres and a standard deviation of 0.13 litres.

    a Find the probability that a randomly selected gear box contains less than the nominal volume.

    It is required by law that no more than 4% of gearboxes contain less than the nominal volume.

    b Find the lowest value of μ that will comply with the law when $\sigma = 0.13$.

    c Find the greatest value of σ that will comply with the law when $\mu = 3.58$ litres.

9 A preserve manufacturer produces a pack consisting of eight assorted pots of differing flavours. The actual weight of each pot may be taken to have an independent normal distribution with mean 52 g and standard deviation σ g. Find the value of σ such that 99% of the packs weigh over 400 g.

10 In a local town election, 2500 people cast a vote for a new mayor. An exit poll predicts that one particular candidate, Antonio Videtta, will secure 38% of the vote.

  a Explain why the binomial distribution is not the best model for modelling this situation.

  b By making appropriate calculations, show that a normal approximation to the binomial distribution can be used in this case.

  c Calculate the mean and variance for the normal approximation.

  d Typically, a candidate needs to secure 40% of the vote to win the election. What is the probability that Antonio Videtta wins the election?

11 A mathematics exam has a mean mark of 75 and a standard deviation of 12. The marks can be assumed to be normally distributed.

  a What is the minimum mark required to be in the top 10% of students sitting the exam?

  b Between which two marks will the middle 50% of students' scores lie?

  c What is the probability that a randomly selected student scores fewer than 40 marks?

12 Duck eggs have a mean weight of 55 grams and a standard deviation of 17 grams. The distribution of their weights may be taken as normal. Eggs weighing less than 42 grams are considered to be small eggs. Eggs weighing more than 42 grams are considered to be standard eggs or large eggs.

  a The farmer selling the eggs would like the proportion of standard eggs to be approximately equal to the number of large eggs. At what weight should the farmer set the division between standard eggs and large eggs?

  b The farmer knows that his particular ducks generally lay green eggs, but that 1% of the eggs are blue. It is considered to be good luck if there are exactly 10 blue duck eggs in a batch of 1000 duck eggs. Using a suitable approximation, find the probability that the farmer finds exactly 10 blue duck eggs in a random batch of 1000 duck eggs.

13 Records from a doctor's surgery show that the probability of waiting more than 15 minutes is 0.0423. The waiting time is normally distributed with a mean of 8 minutes 45 seconds.

  a Calculate the standard deviation.

  b Calculate the probability that the waiting time is less than 5 minutes.

  c The doctors at the surgery typically see 15 000 patients in total each year. Approximately how many of these would be expected to experience a waiting time of less than 5 minutes?

14 In a large college, 56% of students are female and 44% are male. A random sample of 200 students is chosen from the college. Using a suitable approximation, find the probability that more than three-eighths but less than one half of the sample are male.

15 Each cell of a certain plant contains 12 000 genes. A scientist claims that each gene has a 0.003 probability of being duplicated. A cell is chosen at random.

  a Suggest a suitable model for the distribution of the number of duplicated genes in the cell.

  b Find the mean and variance of the number of duplicated genes in the cell.

  c Using a suitable approximation, find the probability that there are exactly 30 duplicated genes in the cell.

16 Staff at a large library suggest that the time spent in the library by a user could be modelled by a normal distribution with mean 75 minutes and standard deviation 15 minutes.

  a Assuming that this model is valid, what is the probability that a user spends:

    i less than an hour in the library

    ii between an hour and an hour and a half in the library?

The library closes at 10.00 pm.

  b Explain why the model suggested by the library staff could not apply to a user who entered the library at 9.00 pm.

  c Estimate an approximate latest time of entry for which the model suggested by the library staff could still be valid.

17 A factory manufactures 300 toy dolls every day. It is known that 2% of dolls are damaged.

  a Using a normal approximation, estimate the probability that at most, four damaged dolls are made in one day.

The quality control system in the factory checks and destroys every damaged doll at the end of the production line. It costs $1.50 to manufacture a doll and the factory sells them for $10.

  b Find the expected profit made by the factory every day.

18 The mass of Chinese leaf cabbages have a normal distribution with mean 500 g and standard deviation 15 g.

  a 10% of Chinese leaf cabbages are lighter than w grams. Find the value of w.

  b Amy picks five Chinese leaf cabbages at random. Find the probability that at least three of them have weights between 505 g and 515 g inclusive.

19 A garden centre sells 12 mixed daffodil bulbs in a bag. It is known that 48% of the bulbs give white flowers. A bag is selected at random.

Calculate the probability that this bag contains:

  a exactly four bulbs with white flowers

  b more than three-quarters of bulbs that give white flowers.

At a plant fair, a bigger bag that contains 50 bulbs is on sale.

c Use a suitable approximation to calculate the probability that a bag of 50 bulbs contains more than 30 bulbs with white flowers.

 20 It is estimated that 7% of the population are left handed. In a random sample of size n, the expected number of people that are left handed is 8.

a Calculate the value of n.

The expected number of people that are left handed in a second random sample is 2.

b Find the standard deviation of the number of people that are left handed in this second sample.

Mathematics in life and work

The mass, X g, of beans put in a tin by machine K is normally distributed with a mean of 300 g and a standard deviation of 11 g.

A tin is selected at random.

1 Find the probability that this tin contains more than 308 g of beans.

The mass stated on the tins is m g.

2 Find m, to 1 decimal place, such that $P(X < m) = 0.02$.

The mass, Y g, of beans put into a cardboard carton is trialled by machine J and is normally distributed with a mean μ g and a standard deviation σ g.

3 Given that $P(X > 290) = 0.95$ and $P(X < 305) = 0.97$, find the values of μ and σ.

SUMMARY REVIEW

Practise the key concepts and apply the skills and knowledge that you have learned in the book with these carefully selected past paper questions supplemented with exam-style questions and extension questions written by the authors.

> Warm-up
> Questions

> A Level
> Questions

> Extension
> Questions

Three Cambridge IGCSE® past paper questions based on prerequisite skills and concepts that are relevant to the main content of this book.

Selected past paper exam questions and exam-style questions on the topics covered in this syllabus component.

Extension questions give you the opportunity to challenge yourself and prepare you for more advanced study.

Warm-up Questions

Reproduced by permission of Cambridge Assessment International Education

1 Chico has a bag of sweets.

He takes a sweet from the bag at random.

The table shows the probabilities of taking each flavour of sweet.

Flavour	lemon	lime	strawberry	blackcurrant	orange
Probability	0.15	0.22		0.18	0.24

i Complete the table. [2]

ii Find the probability that the sweet is lemon or lime.

Answer(b) .... [1]

Cambridge IGCSE Mathematics 0580 Paper 11 Q15 November 2015

2 Paul and Sammy take part in a race.

The probability that Paul wins the race is $\frac{9}{35}$.

The probability that Sammy wins the race is 26%.

Who is more likely to win the race?

Give a reason for your answer. [2]

Cambridge IGCSE Mathematics 0580 Paper 21 Q5 June 2015

3 A company tested 200 light bulbs to find the lifetime, T hours, of each bulb.

The results are shown in the table.

Lifetime (T hours)	Number of bulbs
$0 < T \leqslant 1000$	10
$1000 < T \leqslant 1500$	30
$1500 < T \leqslant 2000$	55
$2000 < T \leqslant 2500$	72
$2500 < T \leqslant 3500$	33

a Calculate an estimate of the mean lifetime for the 200 light bulbs. [4]

b i Complete the cumulative frequency table.

Lifetime (T hours)	$T \leqslant 1000$	$T \leqslant 1500$	$T \leqslant 2000$	$T \leqslant 2500$	$T \leqslant 3500$
Number of bulbs					

[2]

ii Copy the grid and draw a cumulative frequency diagram to show this information.

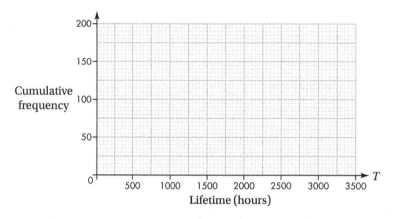

[3]

iii The company says that the average lifetime of a bulb is 2200 hours.

Estimate the number of bulbs that lasted longer than 2200 hours. [2]

c Robert buys one energy saving bulb and one halogen bulb.

The probability that the energy saving bulb lasts longer than 3500 hours is $\frac{9}{10}$.

The probability that the halogen bulb lasts longer than 3500 hours is $\frac{3}{5}$.

Work out the probability that exactly one of the bulbs will last longer than 3500 hours. [4]

Cambridge IGCSE Mathematics 0580 Paper 41 Q6 November 2014

A Level Questions

Reproduced by permission of Cambridge Assessment International Education

1 The lengths, in metres, of cars in a city are normally distributed with mean μ and standard deviation 0.714. The probability that a randomly chosen car has a length more than 3.2 metres and less than μ metres is 0.475. Find μ. [4]

Cambridge International AS & A Level Mathematics 9709 Paper 61 Q1 June 2015

2 In a certain town, 76% of cars are fitted with satellite navigation equipment. A random sample of 11 cars from this town is chosen. Find the probability that fewer than 10 of these cars are fitted with this equipment. [4]

Cambridge International AS & A Level Mathematics 9709 Paper 61 Q1 November 2015

3 Jason throws two fair dice, each with faces numbered 1 to 6. Event A is 'one of the numbers obtained is divisible by 3 and the other number is not divisible by 3'. Event B is 'the product of the two numbers obtained is even'.

i Determine whether events A and B are independent, showing your working. [5]

ii Are events A and B mutually exclusive? Justify your answer. [1]

Cambridge International AS & A Level Mathematics 9709 Paper 61 Q3 June 2015

4 The following back-to-back stem-and-leaf diagram shows the times to load an application on 61 smartphones of type A and 43 smartphones of type B.

	Type A			Type B	
(7)	9766433	2	1358		(4)
(7)	5544222	3	044566667889		(12)
(13)	9988876643220	4	0112368899		(10)
(9)	655432110	5	25669		(5)
(4)	9730	6	1389		(4)
(6)	874410	7	57		(2)
(10)	7666533210	8	1244		(4)
(5)	86555	9	06		(2)

Key: 3 |2| 1 means 0.23 seconds for type A and 0.21 seconds for type B.

i Find the median and quartiles for smartphones of type A. [3]

You are given that the median, lower quartile and upper quartile for smartphones of type B are 0.46 seconds, 0.36 seconds and 0.63 seconds respectively.

ii Represent the data by drawing a pair of box-and-whisker plots in a single diagram on graph paper. [3]

iii Compare the loading times for these two types of smartphone. [1]

Cambridge International AS & A Level Mathematics 9709 Paper 61 Q4 November 2014

5 **a** Find the number of ways in which all nine letters of the word TENNESSEE can be arranged

     i if all the letters E are together, [3]

     ii if the T is at one end and there is an S at the other end. [3]

  b Four letters are selected from the nine letters of the word VENEZUELA. Find the number of possible selections which contain exactly one E. [3]

Cambridge International AS & A Level Mathematics 9709 Paper 61 Q5 November 2015

6 Nadia is very forgetful. Every time she logs in to her online bank she only has a 40% chance of remembering her password correctly. She is allowed 3 unsuccessful attempts on any one day and then the bank will not let her try again until the next day.

  i Draw a fully labelled tree diagram to illustrate this situation. [3]

  ii Let X be the number of unsuccessful attempts Nadia makes on any day that she tries to log in to her bank. Copy and complete the following table to show the probability distribution of X. [4]

x	0	1	2	3
$P(X = x)$		0.24		

  iii Calculate the expected number of unsuccessful attempts made by Nadia on any day that she tries to log in. [2]

Cambridge International AS & A Level Mathematics 9709 Paper 61 Q6 November 2015

7 A typing test is taken by 111 people. The numbers of typing errors they make in the test are summarised in the table below.

Number of typing errors	1–5	6–20	21–35	36–60	61–80
Frequency	24	9	21	15	42

  i Draw a histogram on graph paper to represent this information. [5]

  ii Calculate an estimate of the mean number of typing errors for these 111 people. [3]

  iii State which class contains the lower quartile and which class contains the upper quartile. Hence find the least possible value of the interquartile range. [3]

Cambridge International AS & A Level Mathematics 9709 Paper 61 Q7 June 2014

8 Find the mean and variance of the following data. [3]

    5   –2   12   7   –3   2   –6   4   0   8

Cambridge International AS & A Level Mathematics 9709 Paper 61 Q1 November 2014

9 The table summarises the lengths in centimetres of 104 dragonflies.

Length (cm)	2.0–3.5	3.5–4.5	4.5–5.5	5.5–7.0	7.0–9.0
Frequency	8	25	28	31	12

  i State which class contains the upper quartile. [1]

  ii Draw a histogram, on graph paper, to represent the data. [4]

Cambridge International AS & A Level Mathematics 9709 Paper 61 Q2 June 2015

10 i State three conditions which must be satisfied for a situation to be modelled by a binomial distribution. [2]

George wants to invest some of his monthly salary. He invests a certain amount of this every month for 18 months. For each month there is a probability of 0.25 that he will buy shares in a large company, there is a probability of 0.15 that he will buy shares in a small company and there is a probability of 0.6 that he will invest in a savings account.

ii Find the probability that George will buy shares in a small company in at least 3 of these 18 months. [3]

Cambridge International AS & A Level Mathematics 9709 Paper 61 Q3 June 2014

11 A book club sends 6 paperback and 2 hardback books to Mrs Hunt. She chooses 4 of these books at random to take with her on holiday. The random variable X represents the number of paperback books she chooses.

i Show that the probability that she chooses exactly 2 paperback books is $\frac{3}{14}$. [2]

ii Draw up the probability distribution table for X. [3]

iii You are given that E(X) = 3. Find Var(X). [2]

Cambridge International AS & A Level Mathematics 9709 Paper 61 Q4 June 2014

12 Lengths of a certain type of carrot have a normal distribution with mean 14.2 cm and standard deviation 3.6 cm.

i 8% of carrots are shorter than c cm. Find the value of c. [3]

ii Rebekah picks 7 carrots at random. Find the probability that at least 2 of them have lengths between 15 and 16 cm. [6]

Cambridge International AS & A Level Mathematics 9709 Paper 61 Q5 November 2013

13 A committee of 6 people is to be chosen from 5 men and 8 women. In how many ways can this be done

i if there are more women than men on the committee, [4]

ii if the committee consists of 3 men and 3 women but two particular men refuse to be on the committee together? [3]

One particular committee consists of 5 women and 1 man.

iii In how many different ways can the committee members be arranged in a line if the man is not at either end? [3]

Cambridge International AS & A Level Mathematics 9709 Paper 61 Q7 November 2014

14 James has a fair coin and a fair tetrahedral die with four faces numbered 1, 2, 3, 4. He tosses the coin once and the die twice. The random variable X is defined as follows.
- If the coin shows a **head** then X is the **sum** of the scores on the two throws of the die.
- If the coin shows a **tail** then X is the score on the **first throw** of the die only.

i Explain why $X = 1$ can only be obtained by throwing a tail, and show that P($X = 1$) = $\frac{1}{8}$. [2]

ii Show that P($X = 3$) = $\frac{3}{16}$. [4]

iii Copy and complete the probability distribution table for X. [3]

x	1	2	3	4	5	6	7	8
$P(X = x)$	$\frac{1}{8}$		$\frac{3}{16}$		$\frac{1}{8}$		$\frac{1}{16}$	$\frac{1}{32}$

Event Q is 'James throws a tail'. Event R is 'the value of X is 7'.

iv Determine whether events Q and R are exclusive. Justify your answer. [2]

Cambridge International AS & A Level Mathematics 9709 Paper 61 Q7 November 2013

15 i State three conditions which must be satisfied for a situation to be modelled by a geometric distribution.

Ken chooses a different fruit to eat with his lunch each day. There is a probability of 0.43 that he will choose an apple, there is a probability of 0.28 that he will choose a banana and there is a probability of 0.29 that he will choose an orange.

ii Find the probability that Ken chooses a banana for the first time on the 4th day.

16 It is given that $X \sim N(28.3, 4.5)$. Find the probability that a randomly chosen value of X lies between 25 and 30. [3]

Cambridge International AS & A Level Mathematics 9709 Paper 61 Q1 June 2012

17 Ana meets her friends once every day. For each day the probability that she is early is 0.05 and the probability that she is late is 0.75. Otherwise she is on time.

i Find the probability that she is on time on fewer than 20 of the next 96 days. [5]

ii If she is early there is a probability of 0.7 that she will eat a banana. If she is late she does not eat a banana. If she is on time there is a probability of 0.4 that she will eat a banana. Given that for one particular meeting with friends she does not eat a banana, find the probability that she is on time. [4]

Cambridge International AS & A Level Mathematics 9709 Paper 61 Q6 November 2012

18 A summary of 30 values of x gave the following information:

$$\Sigma(x - c) = 234, \qquad \Sigma(x - c)^2 = 1957.5,$$

where c is a constant.

i Find the standard deviation of these values of x. [2]

ii Given that the mean of these values is 86, find the value of c. [2]

Cambridge International AS & A Level Mathematics 9709 Paper 61 Q1 June 2013

19 Fiona uses her calculator to produce 12 random integers between 7 and 21 inclusive. The random variable X is the number of these 12 integers which are multiples of 5.

i State the distribution of X and give its parameters. [3]

ii Calculate the probability that X is between 3 and 5 inclusive. [3]

Fiona now produces n random integers between 7 and 21 inclusive.

iii Find the least possible value of n if the probability that none of these integers is a multiple of 5 is less than 0.01. [3]

Cambridge International AS & A Level Mathematics 9709 Paper 61 Q5 June 2013

20 Daniel is meeting players from his favourite football team. He knows that 35% of the players have scored a goal this season. How many players should Daniel expect to ask before he finds a player that has scored a goal this season?

21 The lengths of body feathers of a particular species of bird are modelled by a normal distribution. A researcher measures the lengths of a random sample of 600 body feathers from birds of this species and finds that 63 are less than 6 cm long and 155 are more than 12 cm long.

i Find estimates of the mean and standard deviation of the lengths of body feathers of birds of this species. [5]

ii In a random sample of 1000 body feathers from birds of this species, how many would the researcher expect to find with lengths more than 1 standard deviation from the mean. [4]

Cambridge International AS & A Level Mathematics 9709 Paper 61 Q6 June 2012

22 a Seven friends together with their respective partners all meet up for a meal. To commemorate the occasion they arrange for a photograph to be taken of all 14 of them standing in a line.

i How many different arrangements are there if each friend is standing next to his or her partner? [3]

ii How many different arrangements are there if the 7 friends all stand together and the 7 partners all stand together? [2]

b A group of 9 people consists of 2 boys, 3 girls and 4 adults. In how many ways can a team of 4 be chosen if

i both boys are in the team, [1]

ii the adults are either all in the team or all not in the team, [2]

iii at least 2 girls are in the team? [2]

Cambridge International AS & A Level Mathematics 9709 Paper 61 Q7 June 2012

23 The amounts of money, x dollars, that 24 people had in their pockets are summarised by $\Sigma(x-36) = -60$ and $\Sigma(x-36)^2 = 227.76$. Find Σx and Σx^2. [5]

Cambridge International AS & A Level Mathematics 9709 Paper 61 Q2 November 2012

24 A random variable X has the distribution Geo(p). Given that $p = 0.32$, find:

i $P(X = 7)$

ii $P(X < 3)$

iii $P(X > 3)$

25 **i** The random variable Y is normally distributed with positive mean μ and standard deviation $\frac{1}{2}\mu$. Find the probability that a randomly chosen value of Y is negative. **[3]**

ii The weights of bags of rice are normally distributed with mean 2.04 kg and standard deviation σ kg. In a random sample of 8000 such bags, 253 weighed over 2.1 kg. Find the value of σ. **[4]**

Cambridge International AS & A Level Mathematics 9709 Paper 61 Q4 June 2013

26 Lengths of rolls of parcel tape have a normal distribution with mean 75 m, and 15% of the rolls have lengths less than 73 m.

i Find the standard deviation of the lengths. **[3]**

Alison buys 8 rolls of parcel tape.

ii Find the probability that fewer than 3 of these rolls have lengths more than 77 m. **[3]**

Cambridge International AS & A Level Mathematics 9709 Paper 61 Q3 November 2012

27 Box A contains 8 white balls and 2 yellow balls. Box B contains 5 white balls and x yellow balls. A ball is chosen at random from box A and placed in box B. A ball is then chosen at random from box B. The tree diagram below shows the possibilities for the colours of the balls chosen.

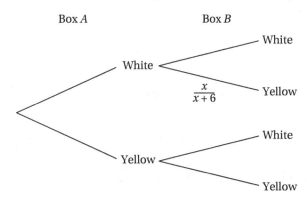

i Justify the probability $\frac{x}{x+6}$ on the tree diagram. **[1]**

ii Copy and complete the tree diagram. **[4]**

iii If the ball chosen from box A is white then the probability that the ball chosen from box B is also white is $\frac{1}{3}$. Show that the value of x is 12. **[2]**

iv Given that the ball chosen from box B is yellow, find the conditional probability that the ball chosen from box A was yellow. **[4]**

Cambridge International AS & A Level Mathematics 9709 Paper 61 Q7 June 2013

28 The lengths of the diagonals in metres of the 9 most popular flat screen TVs and the 9 most popular conventional TVs are shown below.

| Flat screen: | 0.85 | 0.94 | 0.91 | 0.96 | 1.04 | 0.89 | 1.07 | 0.92 | 0.76 |
| Conventional: | 0.69 | 0.65 | 0.85 | 0.77 | 0.74 | 0.67 | 0.71 | 0.86 | 0.75 |

i Represent this information on a back-to-back stem-and-leaf diagram. [4]

ii Find the median and the interquartile range of the lengths of the diagonals of the 9 conventional TVs. [3]

iii Find the mean and standard deviation of the lengths of the diagonals of the 9 flat screen TVs. [2]

Cambridge International AS & A Level Mathematics 9709 Paper 61 Q5 June 2012

29 Chontell is a telephone salesperson. She keeps calling customers until they buy one of her products. The probability that the customer buys one of her products is p, where $0 \leqslant p \leqslant 1$.

i What assumptions are needed to use the geometric distribution?

ii Assuming the assumptions are true, find an expression for the probability that a customer buys one of Chontell's products on the nth phone call.

iii Chontell has found that the probability that a customer buys one of her products on the second phone call is 0.2016. Find the maximum probability of the customer buying one of Chontell's products on the first call.

30 A company set up a display consisting of 20 fireworks. For each firework, the probability that it fails to work is 0.05, independently of other fireworks.

i Find the probability that more than 1 firework fails to work. [3]

The 20 fireworks cost the company $24 each. 450 people pay the company $10 each to watch the display. If more than 1 firework fails to work they get their money back.

ii Calculate the expected profit for the company. [4]

Cambridge International AS & A Level Mathematics 9709 Paper 61 Q5 November 2012

Extension Questions

1 Three boys and two girls stand in a circle in a random order. What is the probability the two girls stand next to each other in the circle?

2 Consider the statement '50% of the doctors scored below average in their final examinations compared to their peers'. Under what circumstances is this statement true if the average referred to is the:

i mean?

ii median?

iii mode?

Justify your answers.

3 An airline company has an aeroplane with a maximum capacity of 250 passengers. However, the airline knows from experience that some passengers miss their flight, so they regularly overbook and sell 275 tickets for the 250 available seats. Data collected by the airline over several flights shows that for this particular flight, there is a 10% chance that the passenger misses their flight.

For a particular flight, what is the probability that more than 250 people arrive on time for the flight?

4 Each of the following statistical diagrams or conclusions could be considered to be misleading. In each case, identify the issue and suggest an improvement.

i

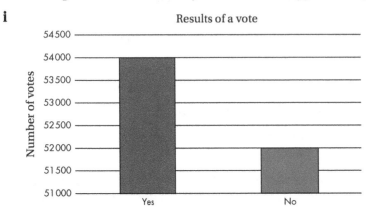

Results of a vote

ii

Mobile phone corporation XXX
Mobile phone sales up – donations to charity down

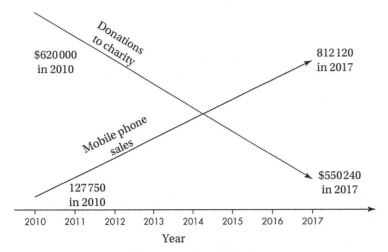

iii

Colour of sweet	Sugar content (g)
red	0.03
orange	0.08
green	0.52
yellow	0.05
blue	0.78

'The average sugar content for one of our sweets is 0.08 g. So, on average, a typical bag of 24 sweets will contain less than 2 g of sugar.'

5 In a group of n people, only one person can be selected to be on the first tourist flight into space. The method for selecting the person is to place n balls in a bag, where 1 ball is green and all of the remaining balls are red. The people are lined up in height order from smallest to tallest. Person 1 (the smallest person) randomly selects a ball from the bag. If it is the green ball, then person 1 is selected. If it is a red ball, then person 1 is not selected. The red ball is not replaced into the bag and the bag is passed to person 2 to repeat the process. This process is repeated with each person in ascending height order and 1 fewer ball each time until the green ball is selected. Show that the probability of being selected is independent of the person's height.

6 How many 5-digit numbers are there that do not contain a 5?

7 Consider a data set with two numbers $a, b \geqslant 0$. Given that the mean of a and b is 2, show that the standard deviation can be given by $\sigma = |a - 2|$.

8 There are 5 people in a room. Assuming that each person has an equal probability of being born on any particular day of the week, find:

i the probability that all 5 people are born on different days of the week

ii the probability that at least two of the 5 people were born on the same day of the week.

9 In a class of 20 maths students, the teacher selects 5 students at random to participate in a maths competition. What is the probability that this group of 5 students includes at least 2 of the top 3 maths students in the class?

10 X is a random variable with the following probability distribution, where $0 \leqslant p \leqslant 1$ and $0 \leqslant q \leqslant 1$.

x	1	2	3	4
$P(X = x)$	p	pq	pq	q

i Given that $E(X) = 2.5$, show that $q = \dfrac{\sqrt{3} - 1}{2}$.

ii Find the value of p.

iii Find the value of $Var(X)$ correct to 3 significant figures.

11 A bag contains some red balls and some green balls and every ball has an equal chance of being selected. There are 2 more green balls than red balls in the bag. A ball is selected at random and then replaced. A second ball is then selected. The probability that the balls are different colours is $\dfrac{195}{392}$.

i Find the minimum number of red and green balls in the bag.

ii Find the probability that 5 randomly selected balls, chosen with replacement, alternate between red and green. Express your answer as a fraction in its simplest form.

12 The random variable $X \sim B\,(n, p)$ where $n \in \mathbb{Z}^+$ and $0 \leqslant p \leqslant 1$.

Show that

$$\frac{P(X = x)}{P(X = x - 1)} = \frac{p(n - x + 1)}{x(1 - p)}$$

where $x \in \mathbb{Z}^+$.

GLOSSARY

bimodal A data distribution that has two modes.

binomial An expression containing exactly two terms, for example, $(x + 2)$.

binomial coefficient The binomial coefficient is
$$\binom{n}{r} = {}^nC_r = \frac{n!}{r!(n-r)!}.$$

binomial distribution A discrete probability distribution in which there are only two outcomes, success and failure, to any trial, the probability, p, of success is the same in each trial and there are a fixed number, n, of independent trials.

box-and-whisker plot Also referred to as a box plot. This is a type of diagram used to display quantitative data. They show the upper and lower quartiles, the median and the maximum and minimum values.

categorical data Data that has been put into groups in some way. For example, eggs are often put into groups depending on their size. Categorical data is qualitative data.

coding Sometimes it may be difficult to work with a set of data due to the values being very large. Subtracting a constant from each member of the data set can make the values much more manageable: this is called coding. Using $y = x + a$ will affect the mean value so that $\bar{y} = \bar{x} + a$, but will not affect the spread of the data so that $\sigma_y = \sigma_x$.

combination A combination is an arrangement of objects chosen from a given set where order does not matter. The number of arrangements of r different objects chosen from a set of n objects where order does not matter is

written as ${}^nC_r$ or $\binom{n}{r}$. The formula
$${}^nC_r = \frac{n!}{r!(n-r)!}.$$

complement The complement of an event is every other possible event. For example, if a dice is rolled and the event A is 'a 6 is uppermost on the dice', then the complement is 'a 6 is not uppermost on the dice' and is written A'. It is important to remember that $P(A) + P(A') = 1$, since it is certain that either event A does happen or event A does not happen.

conditional probability The probability that one event happens given that another event occurs. The probability of event A happening given that event B has happened is written $P(A|B)$.
$$P(A \mid B) = \frac{P(A \cap B)}{P(B)}$$

continuity correction When a discrete probability distribution is approximated by a continuous probability distribution, a continuity correction is needed. For a discrete random variable $P(X = x)$ may have a value other than 0, but for a continuous random variable $P(X = x) = 0$ for all individual values.

continuous data Data that can vary on a continuous scale, i.e. a scale that has no gaps, so it is usually the result of measuring. For example, distance, mass or time. Continuous data is never exact and has to be rounded or grouped in order to be recorded.

cumulative frequency To work out cumulative frequencies you accumulate (add up) the frequencies up to that point. Cumulative frequencies are used to draw cumulative frequency graphs.

These are used to find estimates of quartiles, percentiles and estimates of the number of values above or below a particular value when there is a large amount of data.

data Information obtained from various sources. When you collect information, such as hair colour for each person in a group or the number of cars passing a point at a particular time you are collecting data. There are different types of data, *see* qualitative data, quantitative data and categorical data.

dependent events Events are dependent when the outcome of one affects the outcome of the other, i.e. they are not independent. Events A and B are dependent if $P(A \mid B) \neq P(A)$. Events A and B are dependent if $P(A \mid B) \neq P(A) \times P(B)$.

discrete data Data that can only vary in steps, for example, shoe size, money or the attendance at a tennis match. Discrete data are often counted.

discrete random variable A random variable that can only take specific values, for example, whole number values in the given interval.

event A set of possible outcomes from an experiment, for example, rolling an even number on a die.

expectation The expectation of a random variable X is what you would expect to get if you took a large number of values of X and found their mean. The mean value is called the expected value and is written $E(X)$ or μ. $E(X) = \sum xp(x)$.

factorial If n is a positive integer, then $n!$ is the product of all the positive integers $\leqslant n$ and it is called 'n factorial'. For example 5! $= 5 \times 4 \times 3 \times 2 \times 1 = 120$.

failure   In any trial or experiment a particular event either occurs or it does not occur. When the event does not occur it is described as a failure.

geometric distribution   A discrete probability distribution in which there are only two outcomes, success and failure, to any trial, all trials are independent of each other, the probability, p, of success is the same in each trial and the trials continue until the first success

histogram   A diagram for representing continuous data with rectangles. The areas of any rectangle represents the frequency in that class. To calculate the height of each rectangle the frequency density needs to be found.

$frequency\ density = \dfrac{frequency}{class\ width}.$

Note: the vertical axis is labelled 'frequency density' rather than 'frequency'.

independent events   If events A and B are independent, then the probability of A is unaffected by whether event B happens or not. Events A and B are independent if and only if $P(A|B) = P(A)$. Events A and B are independent if and only if $P(A \cap B) = P(A) \times P(B)$. Remember that $P(A \cap B)$ is the probability that both events A and B occur.

interquartile range   The interquartile range is a measure of variation. It is the difference between the upper quartile and lower quartile, i.e. $Q_3 - Q_1$.

lower quartile   For a set of ordered data the median divides the set into two 'equal' parts. The lower quartile, Q_1, divides the lower half into two 'equal' parts.

mean   The mean, ($\overline{x}$), is an example of an average. It is the sum of all the values divided by the total number of values. For a simple set of n numbers the mean is $\overline{x} = \dfrac{\Sigma x}{n}$, and for a frequency distribution the mean is $\overline{x} = \dfrac{\Sigma fx}{\Sigma f}$.

median   The median is an example of an average. It is found by listing all the values in numerical order and choosing the value in the middle. If there are an even number of values then the median is the mean of the middle pair of values.

mode   The mode is an example of an average. It is the value that occurs most frequently. Sometimes a data set may have no mode or more than one mode. The mode is sometimes called the modal value. For grouped data the modal group or modal class may be found.

mutually exclusive events   Two or more events that cannot both occur at the same time. For example, an ordinary die can be rolled, if event A is 'an even number is uppermost' and event B is 'an odd number is uppermost' then events A and B are mutually exclusive, the number uppermost cannot be both odd and even. This means that $P(A \cap B) = 0$ and $P(A \cup B) = P(A) + P(B)$.

normal distribution   Normal distributions occur in nature. They are the probability distribution for a continuous random variable, for example, the length of leaves for a particular type of tree. Values that are closer to the mean value have a higher probability than values further away from the mean value. When represented graphically, normal distributions have a bell-shaped curve, they are symmetrical about the mean and the total area under the curve is 1.

numerical data   Another way to describe quantitative data. Numerical data only takes values that are numbers.

outlier   Any data value that does not fit in well with the rest of the data set. Outliers are often defined to be values that are less than (Q1 – 1.5 × IQR) or more than (Q3 + 1.5 × IQR).

percentile   The values that divide a rank-ordered set of elements into 100 'equal' parts in the same way that quartiles divide a data set into 4 'equal' parts.

permutation   A permutation is an arrangement of objects chosen from a given set where order matters. The number of arrangements of r different objects chosen from a set of n objects where order matters is written as ^nP_r. The formula is $^nP_r = \dfrac{n!}{(n-r)!}.$

probability   The measure of likelihood that an event will happen. The probability of an event A happening is written $P(A)$. $P(A)$ is always between 0 and 1. If $P(A) = 0$ then the event A will not happen. If $P(A) = 1$ then the event A is certain to happen. Probability of an event = (number of successful outcomes) / (total number of equally likely outcomes).

probability distribution   The set of all possible values a random variable can take, with the associated probability of each value occurring. For a discrete probability distribution this will be in the form of a table.

qualitative data   Qualitative data expresses a quality of something, such as taste, colour, smell or touch. It is not numerical.

quantitative data   Quantitative data is always numerical (a number). Quantitative is similar to the word 'quantity', an amount. There are two types of quantitative data, discrete data and continuous data.

quartile   Quartiles (Q_1, Q_2, Q_3) divide a set of ordered data into four parts (quarters), such that each part has the same number of data values. In the same way that the median, Q_2, divides the whole data set into two 'equal' halves, the lower quartile, Q_1, divides the lower half into two 'equal' parts and the upper quartile, Q_3, divides the upper half into two 'equal' parts.

random sample A sample is part of a population. A random sample is a sample that has been obtained from the population in such a way that each member of the population has an equal probability of being included in the sample and the selection has been made using a random method, such as using random number tables or a random number generator.

random variable A variable is something that can take a value which varies. In statistics this is either a measurement or an observation from a trial or an experiment. When the value may be subject to some random variation, the variable is described as a random variable. Random variables are usually denoted by an upper case letter and the values taken by a random variable are usually denoted by lowercase letters. For example, the probability the random variable X takes a particular value x is usually written as P($X = x$). It can also be written as p(x).

range The range is a measure of variation. It is the difference between the highest and lowest values in the data set.

sample space A sample space diagram shows all the possible outcomes. They are useful when a trial consists of two separate things, such as flipping a coin and throwing a die.

standard deviation Standard deviation, σ, is used as a measure of variation. It is the square root of the variance. $\sigma = \sqrt{\sum \frac{(x - \bar{x})^2}{n}}$ and $\sigma = \sqrt{\frac{\sum x^2}{n} - \bar{x}^2}$ are equivalent formulae for finding the standard deviation.

standardised normal distribution The normal distribution that has a mean value of 0 and a standard deviation of 1 is called the standardised normal distribution. The continuous random variable for a standard normal distribution is denoted by the letter Z, and the values it takes by z; we write $Z \sim N(0,1)$

standardising The process of converting any normal distribution with mean μ and standard deviation σ to a standardised normal distribution with mean 0 and standard deviation 1 is called standardising. To standardise $X \sim N(\mu, \sigma^2)$ to $Z \sim N(0,1)$ use the formula $z = \frac{value - mean}{standard\ deviation} = \frac{x - \mu}{\sigma}$ Standardising allows the table of values to be used to find probabilities for any normal distribution.

stem-and-leaf diagram Diagram that allows two sets of data to be displayed side by side on the same stem. Leaves must be single digits and ordered. A key must also be provided.

success In any trial or experiment a particular event either occurs or it does not occur. When the event occurs it is described as a 'success'.

trial A trial is a single experiment, for example, when a die is thrown each throw is called a trial.

uniform distribution A probability distribution in which each value is equally likely, for example, the throwing of a fair die.

upper quartile For a set of ordered data the median divides the set into two 'equal' parts. The upper quartile, Q_3, divides the upper half into two 'equal' parts.

variance Variance, σ^2, is used as a measure of variation. The deviation (difference) of each value from the mean is found, squared, and then the mean of these squared deviations is found. $\sigma^2 = \sqrt{\sum \frac{(x - \bar{x})^2}{n}}$ and $\sigma^2 = \sqrt{\frac{\sum x^2}{n} - \bar{x}^2}$ are equivalent formulae for finding the variance. For frequency distributions the formulae are $\sigma^2 = \sqrt{\frac{\sum f(x - \bar{x})^2}{\sum f}}$ and $\sigma^2 = \sqrt{\frac{\sum f x^2}{\sum f} - \bar{x}^2}$.

INDEX